W9-CGW-007

101 THINGS TO DO WITH YOUR PRIVATE PILOT'S LICENSE

101 THINGS TO DO WITH YOUR PRIVATE PILOT'S LICENSE

LeRoy Cook

THIRD EDITION

McGraw-Hill
New York • Chicago • San Francisco • Lisbon • London • Madrid
Mexico City • Milan • New Delhi • San Juan • Seoul
Singapore • Sydney • Toronto

The McGraw·Hill Companies

Cataloging-in-Publication Data is on file with the Library of Congress

Copyright © 2004, 1990, 1985 by The McGraw-Hill Companies, Inc. All rights reserved. Printed in the United States of America. Except as permitted under the United States Copyright Act of 1976, no part of this publication may be reproduced or distributed in any form or by any means, or stored in a data base or retrieval system, without the prior written permission of the publisher.

1 2 3 4 5 6 7 8 9 0   DOC/DOC   0 9 8 7 6 5 4 3

ISBN 0-07-142258-7

The sponsoring editor for this book was Larry S. Hager, the editing supervisor was Stephen M. Smith, and the production supervisor was Sherri Souffrance. It was set in Fairfield Medium following the NBF design by Kim Sheran and Deirdre Sheean of McGraw-Hill Professional's Hightstown, N.J., composition unit. The art director for the cover was Anthony Landi.

Printed and bound by RR Donnelley.

This book is printed on recycled, acid-free paper containing a minimum of 50% recycled, de-inked fiber.

McGraw-Hill books are available at special quantity discounts to use as premiums and sales promotions, or for use in corporate training programs. For more information, please write to the Director of Special Sales, McGraw-Hill Professional, Two Penn Plaza, New York, NY 10121-2298. Or contact your local bookstore.

Information contained in this work has been obtained by The McGraw-Hill Companies, Inc. ("McGraw-Hill") from sources believed to be reliable. However, neither McGraw-Hill nor its authors guarantee the accuracy or completeness of any information published herein and neither McGraw-Hill nor its authors shall be responsible for any errors, omissions, or damages arising out of use of this information. This work is published with the understanding that McGraw-Hill and its authors are supplying information but are not attempting to render engineering or other professional services. If such services are required, the assistance of an appropriate professional should be sought.

To Dennis Shattuck, whose unfailing tolerance and support has meant so much throughout our association.

CONTENTS

Introduction   *xi*
The 101 Things   *xiii*

PART ONE. STRIKING OUT ON YOUR OWN

CHAPTER ONE. A LICENSE TO LEARN   3

CHAPTER TWO. GOING SOMEWHERE?   7

CHAPTER THREE. EFFICIENT FLIGHT PLANNING   13

CHAPTER FOUR. THE FLYING VACATION   19

CHAPTER FIVE. LOW-LEVEL FLYING   23

CHAPTER SIX. DODGING TALL TOWERS   29

CHAPTER SEVEN. THE VFR FLIGHT PLAN   33

CHAPTER EIGHT. USING GPS IN THE COCKPIT   37

PART TWO. TRAFFIC AND AIRPORTS

CHAPTER NINE. MIDAIR MENACE   43

CHAPTER TEN. UNCONTROLLED AIRPORTS   47

CHAPTER ELEVEN. GRASS STRIPS   53

CHAPTER TWELVE. TOWER-CONTROLLED AIRPORTS   59

CHAPTER THIRTEEN. CLASS CHARLIE AIRSPACE   65

CHAPTER FOURTEEN. CLASS BRAVO AIRSPACE   69

PART THREE. WEATHER

CHAPTER FIFTEEN. PREFLIGHT BRIEFING   79

CHAPTER SIXTEEN. PERSONAL FORECASTING   85

CHAPTER SEVENTEEN. COPING WITH MARGINAL VFR   91

CHAPTER EIGHTEEN. TRAPPED IN IFR WEATHER   97

PART FOUR. SURVIVING THE SEASONS

CHAPTER NINETEEN. WIND WISDOM   105

CHAPTER TWENTY. THUNDERSTORMS: SUMMER MONSTERS   111

CHAPTER TWENTY-ONE. SQUALL LINE ENCOUNTER   117

CHAPTER TWENTY-TWO. MANAGING HEAT STRESS   123

CHAPTER TWENTY-THREE. HAZY SUMMER DAYS   127

CHAPTER TWENTY-FOUR. FOG, INSIDIOUS ENEMY   131

CHAPTER TWENTY-FIVE. COLD, COLD START   135

CHAPTER TWENTY-SIX. CHILLER ICE   141

CHAPTER TWENTY-SEVEN. FLYING SNOW   145

Part Five. Using Your Head   151

CHAPTER TWENTY-EIGHT. GO/NO-GO JUDGMENT   153

CHAPTER TWENTY-NINE. SLOW DOWN TO
SAVE TIME   157

CHAPTER THIRTY. STRESS MANAGEMENT   161

CHAPTER THIRTY-ONE. CONSIDERATE
OPERATIONS   165

CHAPTER THIRTY-TWO. SENSIBILITY   171

CHAPTER THIRTY-THREE. CURRENCY   177

Part Six. Your Own Airplane

CHAPTER THIRTY-FOUR. FLYING CLUBS   185

CHAPTER THIRTY-FIVE. FIRST LOVE   189

CHAPTER THIRTY-SIX. CUSTOMIZED CHECKLIST   193

CHAPTER THIRTY-SEVEN. TEST FLYING AN
AIRPLANE   201

CHAPTER THIRTY-EIGHT. THOROUGH
POSTFLIGHT INSPECTION   207

CHAPTER THIRTY-NINE. ACCIDENTS   213

PART SEVEN. GOING ON

CHAPTER FORTY. FLIGHT REVIEW DUE?   219

CHAPTER FORTY-ONE. AEROBATICS   225

CHAPTER FORTY-TWO. TAMING A TAILWHEEL   229

CHAPTER FORTY-THREE. HIGH-PERFORMANCE
CHECKOUT   237

CHAPTER FORTY-FOUR. FLYING FOR MONEY   243

CHAPTER FORTY-FIVE. INSTRUMENT RATING—
WORTHWHILE?   247

CHAPTER FORTY-SIX. IFR ITINERARY   251

CHAPTER FORTY-SEVEN. FILE IFR OR STAY VFR?   255

CHAPTER FORTY-EIGHT. REGAINING IFR
CURRENCY   259

CHAPTER FORTY-NINE. SPLASHING IT ON   263

CHAPTER FIFTY. SOARING SPIRITS   269

CHAPTER FIFTY-ONE. UPLIFTING EXPERIENCE   275

CHAPTER FIFTY-TWO. MULTIENGINE RATING   281

CHAPTER FIFTY-THREE. ATP CERTIFICATE   287

Index   291

INTRODUCTION

This book is for young pilots—young in hours of flying experience, if not in age. When a new pilot finishes the course of training leading to the private pilot's certificate, he or she often enters a letdown phase, during which it is natural to wonder "What do I do now?" No longer carefully shepherded through each hour, the pilot is suddenly left to seek his or her destiny unaided.

This book will attempt to pick up where student training leaves off, taking the new pilot into areas only lightly touched in training—or perhaps omitted altogether. We will explore those first trips, enter unfamiliar airports, negotiate for FSS and ATC services, and learn more about weather. We'll talk about each of the flying seasons, from breezy spring to hazy summer, and on into foggy autumn and frigid winter.

Most important, we're going to discuss the pilot's need for understanding his or her limitations by developing the judgment and careful attitudes that will prevent a bad experience from turning into a tragedy. When and how to buy and fly that first airplane will be covered, and we'll also talk about going on, taking advanced training for skills and licenses to be added to those of the basic private pilot certificate.

Learning to fly should continue throughout a pilot's career, an unending process more properly termed "learning about flying." When a beginning student asks me "How long does it take to learn to fly?" I always employ my favorite rejoinder: "I don't know—I've never finished."

Each flight is a chance to learn more, and with all the wide and varied experiences available to the modern aviator there should be no reason to grow stale.

Come explore with us.

LeRoy Cook

THE 101 THINGS

Each of these may be found in the text on the given page; the words that mention or introduce a discussion of the topic on the page are set in **boldface.**

1. Take passengers for rides (p. 4)
2. Keep on improving your skills (p. 4)
3. Learn to flight plan trips quickly (p. 10)
4. Flight plan for efficiency (p. 13)
5. Learn how to lean (p. 14)
6. Take a flying vacation (p. 19)
7. Learn to be flexible (p. 22)
8. Flying at low level (p. 23)
9. Navigating down low (p. 24)
10. Avoiding tall towers (p. 30)
11. Using the VFR flight plan (p. 33)
12. Using aviation GPS (p. 38)
13. Where to look for other planes (p. 43)
14. Ways to avoid a midair collision (p. 45)
15. Entering uncontrolled patterns (p. 49)
16. Courtesy on the party line (p. 51)
17. How to operate from a grass surface (p. 54)
18. Assessing the worth of turf (p. 55)
19. Arriving at a tower-controlled field (p. 60)
20. Departing from a Class D airport (p. 63)
21. Arriving at a Class C airport (p. 65)
22. Departing from Class C fields (p. 67)
23. Entering Class B airspace (p. 71)

24. Departing a Class B airport (p. 74)

25. Prebrief yourself on weather (p. 79)

26. Analyzing the weather briefing (p. 83)

27. Personalizing weather (p. 86)

28. Eyeballing the weather (p. 87)

29. Surviving marginal weather (p. 92)

30. When to say no-go (p. 94)

31. Escaping from instrument weather (p. 97)

32. Emergency letdowns (p. 100)

33. Defining too much wind (p. 105)

34. Out-thinking the wind (p. 108)

35. Dealing with thunderstorms (p. 111)

36. In-flight thunderstorm encounters (p. 115)

37. Squall line techniques (p. 118)

38. Taking refuge from storms (p. 120)

39. Preparing the airplane for hot weather (p. 124)

40. Preparing the pilot for hot weather (p. 126)

41. How to handle haze (p. 127)

42. What lurks in haze (p. 130)

43. How to fly fog (p. 131)

44. Where fog comes from (p. 133)

45. Getting going in the cold (p. 135)

46. Keeping the engine running in the cold (p. 138)

47. Don't go where there's ice (p. 141)

48. Ice under VFR (p. 143)

49. Assessing snow cover from aloft (p. 146)

50. Departing a snow-covered runway (p. 147)

51. The importance of honesty (p. 153)

52. Determining limits (p. 154)

53. Slow down for safety (p. 157)

54. Work with the plane (p. 159)

55. Unloading stress on the ground (p. 162)

56. Avoiding stress in the air (p. 164)

57. Departing courteously (p. 165)

58. Arriving with consideration (p. 168)
59. Learning why not (p. 171)
60. No booze when flying (p. 174)
61. Checking yourself out (p. 178)
62. When are you current? (p. 181)
63. Sharing an airplane successfully (p. 185)
64. Staying financially sound (p. 188)
65. Buying your first plane (p. 189)
66. Insurance considerations (p. 190)
67. Using checklists (p. 193)
68. Making up a checklist (p. 194)
69. Returning a plane to service (p. 201)
70. Test flying a rebuilt plane (p. 203)
71. Reporting postflight squawks (p. 208)
72. Leaving your plane ready to fly (p. 210)
73. What is an accident? (p. 213)
74. Who do you call when you have an accident? (p. 214)
75. Why take a flight review? (p. 220)
76. Find the right instructor (p. 222)
77. Approach aerobatics cautiously (p. 226)
78. Aerobatics the wrong way (p. 227)
79. Why the tailwheel? (p. 229)
80. Wheel landing tips (p. 234)
81. What's high performance? (p. 237)
82. Checking out in bigger planes (p. 239)
83. Do you need a commercial ticket? (p. 244)
84. What's on a commercial checkride? (p. 245)
85. How to get the instrument rating (p. 248)
86. Is the instrument rating worthwhile? (p. 249)
87. The first solo IFR (p. 251)
88. When IFR is not recommended (p. 254)
89. Is it IFR or VFR? (p. 255)
90. Staying down under safely (p. 257)
91. Regain IFR skills (p. 260)

92. Simulate failures (p. 262)
93. How to gain water wings (p. 263)
94. Taxiing on water (p. 265)
95. Docking technique (p. 268)
96. Soaring away (p. 269)
97. Getting certified for gliders (p. 273)
98. Helicopter differences (p. 276)
99. Flying helicopters (p. 278)
100. Flying multiengine airplanes (p. 282)
101. The airline transport rating (p. 287)

STRIKING OUT
ON YOUR OWN

A LICENSE TO LEARN

It's a good feeling to have the private pilot checkride passed, to have all that dual and solo practice behind you. Now you're free to go out and just fly when and where you want, and with anybody you can talk into going along. Welcome, new pilot, to the real world of aviation.

Do you know what you've just acquired? A license to *learn*, that's what. Let's face it, you aren't a bit safer or smarter than you were before you passed your checkride, yet previously you couldn't have taken me for a ride, and now you can. The difference between then and now is that little slip of paper that says "Private Pilot" on it; you've been tested and found free of unsafe gaps in skill and knowledge. You've got gaps all right, it's just that the government feels they are inconsequential enough to be filled in while you engage in your own personal flying.

A new private pilot proudly shows off his temporary certificate and receives congratulations from his flight instructor.

Never, *ever*, stop learning about flying if you want to be around to give your grandchildren airplane rides and to eventually pass away of natural causes. There is so much to know I rather doubt that anyone can lay claim to all of it, yet you will look back on this moment years from now and truly realize how little you knew when you became a private pilot. You've been given all the training the average student can afford; the rest just has to come later.

THE FIRST PASSENGERS

You've probably got a long list of people you have been promising to take for a ride, so call 'em up as the opportunity arises and share your joy. But, please, do aviation a favor and pick a good, quiet, still-air hour for their ride if they haven't been up before. Treat them gently; explain what you're doing so they won't jump and clutch when the wings bank and the sound of the engine changes. Keep the turns gentle and the climbs and descents shallow; don't try to prove your prowess as a fighter pilot.

Some people may seem reluctant to ride with you, a little afraid for their necks, perhaps, because they're being flown by a newly rated pilot. If they would only read the accident statistics, they would find that you're a safer bet now than you will be a couple of hundred hours down the road. Right now you're still cautious and unsure of yourself. You'll ask for advice, you'll use your checklist, you'll preflight carefully. Sadly, all this tends to change when your logbook reaches the vicinity of the 200-hour mark. With a couple of hundred hours of flying time, you're no longer a green hand, you're feeling like an old experienced pilot. You don't need those student pilot crutches anymore; you figure you've been around and seen it all. Most 200-hour pilots make it through this settling period, but some don't. The accident charts show a similar trend around the magic 1,000-hour mark. "This is a lot of flying time," you'll think, "surely I know it all by now." Take it from me—*you don't.* I'm still learning just as much today as when I passed that thousandth hour.

NEVER STOP GETTING BETTER

Now, **where you go from here is up to you.** You can fly the next 500 hours and gain 500 hours of experience, or you can log 500 hours and get 1 hour's experience repeated 500 times. Take your choice: Either learn from each hour and get better, or sit there insensitive and regress. Right now, you're probably thinking, "Heck, I'll bet some of the private pilots I know couldn't pass that flight test." You're right—they stopped learning the day they passed their checkride. They have never gone on to master 30-knot crosswinds or high-density traffic; they're right there where they were as student pilots. Resolve not to let this happen to you.

You told your instructor you would be back every little bit for some refresher training. Did you notice his or her half-smile? They've heard every pilot that's graduated make that statement, and it almost never happens. Please, surprise them by coming back. As you will find out in the coming years, a short flight review does not constitute adequate refresher training. In keeping with your desire to learn all you can, get curious about something once in a while, and take an hour of dual to see what it's all about. Maybe you want to see inside a cloud, for real; get a certified flight instructor, instrument (CFII) and try it—the right way. Maybe you want to see the world roll around the airplane; if so, take a sample aerobatic lesson. We all need a CFI to ride with us now and then, so find some excuse to make it interesting and you'll be more likely to do it.

Convinced that you want to get sharp? Good, just keep your eyes and ears open and fly—that's the way to begin. Now that you're a real pilot, take a short weekend cross-country trip or two. Just avoid a rigid schedule so the weather can't trap you and have more than one destination in mind so you can outflank a front. Get out there and see how it really is. If you stay in the local area, hopping friends on a Sunday afternoon, you'll gradually lose your confidence and desire. Besides, someday you'll want to see another seacoast or the other side of the mountains, and you need to warm up first by making small trips before tackling a week-long journey.

BATTLING THE BUDGET

Can't afford it, you say? Surprise, none of us can. Most of us do without something else to support a flying habit—things like lunch, golf, or a new car. If you can't fly as much as you want—and who can—at least hang around the airport and keep your antenna up, receiving the vibes. It'll keep you out of the bars, anyway, and that'll save money for flying later. Read all those flying magazines so you can benefit from the experiences of the other guys and gals; it'll all be helpful someday.

Thinking about buying an airplane? This is not the time; if you have the money available, somebody may sell you something you don't really need. You should rent the various types you're interested in, if possible, or maybe offer to pay expenses for an extended demonstration. Don't buy something because it's pretty, or after only one hop around the patch. Take it out and fly it cross-country an hour or two; that short jaunt may save you much more than it'll ever cost you. Go to a trusted fixed-base operator (FBO), CFI, or airframe and power plant mechanic (A&P), and ask what he or she thinks; pay for the opinion if necessary, but don't buy an airplane in haste.

On the other hand, you might as well give up and buy something that isn't exactly perfect as soon as you can make up your mind, just so you can maintain proficiency at your convenience. If you can make a good rental deal on a little-used airplane, fine and dandy, but after you are forced to cancel out a few trips and drive 200 miles in bright sunshine, you'll probably be an airplane sales prospect.

You might think weather is the great bugaboo of this business, and you'd be right. It turns up in the accident reports all too often, more than any other single factor, and it behooves you to hone and sharpen your weather sense. Whether you're flying or not, look up at the sky every day and analyze what you see there. Know what various types of clouds mean, which way good weather lies, and when a forecast isn't reliable. You must learn to be your own weatherperson; don't abdicate this responsibility to others.

As time goes on, you will someday be looked upon as an old, knowledgeable pilot. Use the coming years and hours wisely so that this assessment will not be a mistaken one.

GOING
SOMEWHERE?

When you suddenly have a trip to a place you've never heard of, getting organized to leap off in a hurry takes a bit of doing. Spreading out a chart and searching for an obscure destination leads only to fruitless frustration and, likely as not, asking other pilots for a clue will provide the standard response, "Never heard of it." Fortunately, there is a better way to plan a flight, involving some diverse but successful methods.

FLIGHT PLANNING

The term "flight planning" means looking over the route, laying out a course, and checking into fuel requirements and alternate airports. All of these assume that you know where you're heading and if you don't, reach for the road atlas that ought to be in every flight case. No, we're not going to use it for navigation, but the atlas does make it easy to locate an obscure town, so long as you know the state, by consulting an alphabetical index in the back of the book where each town is listed with map coordinates. It beats the heck out of looking over a sectional with a magnifying glass, only to find that the place you've been hunting was an inch beyond the edge of the chart.

Having used the road atlas to pin down this fishing spot, relatives' hometown, or sales prospect's plant site, note its approximate relationship to a large city (such as 40 miles south of Cincinnati on a major highway) and consult the appropriate sectional chart for the nearest airport. If you don't have the chart at hand, a diagram in the legend panel of every sectional will tell you which chart covers the approximate area in question.

Choosing a destination airport is difficult with the meager information displayed on the chart, making it wise to consult the *Airport/Facility Directory* for such details as runway composition, obstructions, hours of operation, and the like. More data are found in the annually published

In addition to the ubiquitous sectional chart, supplemental information from the *Airport/Facility Directory,* AOPA Airport Directory, and *Flight Guide Airport Manual* is essential for planning a flight.

airport directory from the Aircraft Owners and Pilots Association. The AOPA book compiles information on just about every airport in the United States, including such choice tidbits as the brand of fuel sold, the operator's phone number, motel names and numbers, type of aircraft dealership—all kinds of good stuff. Although the AOPA directory isn't the official word, being issued only every other year, it is helpful. Other airport directories are published, notably *Flight Guide* and *Jeppesen's JeppGuide,* and various state aviation departments issue state directories, but most are limited in some fashion to keep down the size of the document, such as showing only data for paved airports or for those with runways over a specified length. Be sure your information is current; don't be above making a phone call if your airplane is going to require special services or runway dimensions. In the absence of such information, plan to have fuel in reserve to divert to an alternate airport.

EYEBALL THE ROUTE

Now eyeball the route. If the trip is a long one, laying it out can take up the entire living room floor. And, because the sectional charts have two sides, it may not be possible to see all of a north-south route at once. For fast guesstimating, invest in an instrument flight rules/visual flight rules (IFR/VFR) planning chart, a 35- by 36-inch map that covers the whole United States at 47 nautical miles to an inch. It shows each very high frequency omnidirectional radio range (VOR) station and the airways between, making it mighty handy to see which airways take you where you want to go, as well as where practical fuel stops might be made. Instrument pilots may use Jeppesen's low-altitude planning charts for similar purposes.

Long trips are perhaps most easily tackled on the VOR airways, where there is helpful course guidance preplotted and a guarantee of radio reception. But you may want to fly RNAV-direct by drawing a pencil line straight across the world; if so, break an extralong flight into segments terminating in pit stops every 2 or 3 hours. Solo trips can be stretched out, of course, but don't abuse a passenger's endurance with optimistic flight planning. A yardstick or the straightedge of another chart will serve to rule off the course.

Flight planning software for your PC simplifies the layout procedure, but review it carefully to make sure it takes you where you want to go. And most aviation GPS (global positioning system) receivers will generate a direct route for you without dragging out the chart, a tempting but unwise method that can lead you into areas you shouldn't visit. Check the route for hazards on your charts. You will wish to avoid busy traffic areas around major airports, military operations areas, and restricted areas. Also look for sparsely populated regions that have few airports; you'll want to alter your route slightly to stay closer to civilization. High terrain also may preclude a direct flight, if an airway would lead to your destination only by climbing to oxygen-required altitudes.

Always seek a route with good alternatives in case you don't find ideal weather on your flight. It's wise to have two routes in mind, one for good VFR weather and one for marginal VFR use. If the weather goes sour, you may wish to abandon the VOR airway across trackless wasteland in favor of following a friendly highway or railroad that has towns and airports every few miles. IFR pilots should take a minute to consider a low-level alternate route in case it's necessary to go underneath the weather because of embedded thunderstorms or icing conditions. When flight conditions deteriorate, it's nice to be able to swing over to the preplanned, alternate low-weather route and press on with a line of position under the wings and an alternate airport a few miles away at all times.

The old road map comes into play once again as a points-of-interest guide for the route. Part of flying's charm is the ability to look down on monuments, landmarks, and historical places; many of these are noted in the road. You may be on a business trip, but that doesn't mean you can't make a 5-mile detour to see Monticello from the air. Naturally, you'll stay 2,000 feet above the ground to avoid brushing the groundlings' sensibilities.

WRITING IT DOWN

For pity's sake, write down some of this hard-won information as you dig it out. Your flight log form may consist of a yellow legal pad, which is fine if you can make it do what you want, but you should have mileages, headings, and frequencies listed in order of use, with appropriate remarks such as "High terrain ahead" or "Watch for Harper's Ferry to left." Having mileages premeasured helps calculate "howgozit" estimates in flight, when headwinds cut into fuel supplies or an unexpected tailwind tempts you to skip a stop.

If you're the type who hates to waste time, even while cruising cross-country, play around with plotting aircraft performance. You can time descents and climbs, calculate true airspeed (TAS), groundspeeds, ETAs, and fuel burns from various tanks, all in the interest of learning more about your airplane's capabilities. Or your tastes may run to meteorology, keeping track of temperature lapse rates and frontal zones encountered, and indulging in pressure-pattern flying for optimum block speeds. Perhaps aerial photography intrigues you. The reason I mention these diversions in a discourse on flight planning is that I invariably forget about the camera or calculator—or even a spare pencil—until I'm airborne, so put them in the flight case *now*, before the balloon goes up.

GETTING ORGANIZED IN A HURRY

You won't always have the chance to indulge in elaborate flight planning. As a charter pilot, I have often made departures within 30 minutes after a phone call, but there is always time for *some* preplanning. At least take a minute to mark and measure the route, something that is next to impossible in the confines of the cockpit. If you're getting off VFR with a possibility of running into IFR weather, find the logical points at which to plug into the system for a clearance if you're instrument rated and current. You can always jot down times over a specific area and figure estimates later, on the face of the chart if need be, but do your basic plotting down on terra firma, where 5 minutes' work can be time well spent.

Now assemble your flight planning goodies in a ditty bag or whatever, so it will all leave the house with you when you might not have full command of your faculties. You don't have to tote a 40-pound flight case, just a clipboard and a couple of pencils in a plastic pouch, but it's awfully embarrassing to reach for a chart and find it was the one that slipped under the table last night.

As a final flight planning tip, take along an RON (remain overnight) kit even on 1-day flights. Having the toothbrush along ensures that you will probably get home tonight; the first time you forget to carry your kit will be the day the forecasts go sour and you'll have to stay out with no razor or spare socks. An RON kit is like a good luck charm—as any charter pilot knows.

As the old saw goes, "Plan your flight and fly your plan." It's still good advice to follow.

EFFICIENT FLIGHT PLANNING

Would you like to fly more and pay less? No, I'm not advocating a switch to an ultralight vehicle, I'm talking about getting more efficiency from your regular bird. Most of us have let poor flight planning and sloppy cruise-control technique become a habit. With the increased costs of avgas, we need to investigate every possible means of trimming our fuel consumption figures. It isn't necessary to scrutinize the upper-air charts or invest in complex engine analyzers to do this either. A few simple steps will stretch your fuel budget to cover 10 to 20 percent more mileage.

FLY MORE AND PAY LESS

Start by renewing your acquaintance with the airplane's pilot's operating handbook. Every airplane/powerplant combination has a most efficient power setting for each flight condition, and you should no longer be content to use one set of numbers for all conditions. Suppose, for instance, you're just out tooling around the local area with some friends. Why blast about at 75 percent power, with the attendant noise and dollar drain, when you're really not going anywhere? A 60 percent setting will serve just as well, and in larger airplanes you may want to cut back to 50 percent or so. By doing so, you'll be saving fuel for the days when you really *are* in a hurry.

FINDING THE AIRPLANE'S BEST ALTITUDES

A few general facts will become evident after reviewing the cruise charts. First, significant savings will result from using 60 or 65 percent power settings rather than 70 or 75 percent. For example, a Cardinal RG at 7,500 feet will cruise at 169 mph (miles per hour) on 74 percent power, burning 10.6 gallons per hour (gph). Yet at 60 percent power it will still do 155 mph and use only 8.7 gph. By giving up 14 mph, one saves nearly 2 gph; the miles per gallon went from 15.9 to 17.8. Second, it pays to fly high. The Cardinal RG gives 149 mph on 60 percent power at 2,500 feet, but 159 mph can be had on the same 60 percent at 10,000 feet. That's 10 mph extra without any increase in cruise fuel flow.

Now, we all know it isn't practical to cruise at 10,000 feet on every hop, even if weather conditions would permit it. It takes time and fuel to get up there, so the trade-off must become advantageous before it will pay to go high. The fuel required to take the Cardinal RG from 5,000 to 10,000 feet amounts to 2 gallons; therefore, a flight duration of more than 2 hours would be required to save enough fuel to warrant the extra climb. As a rule of thumb, you can usually afford to climb as high as 5,000 feet above the starting point on trips of 100 miles, and as high as 10,000 feet for 200-mile hops.

A turbocharger will allow even higher and more efficient flight if your flights are long enough and you don't mind wearing an oxygen mask. Climb rates will remain brisk instead of tapering off above 10,000 feet, and fuel flow remains constant while cruising faster in the thin air. There's no such thing as a free lunch, though, and the cost and upkeep of a turbo, as well as its inherent slight increase in fuel burn, must be weighed against the number of high-altitude trips you will be making.

A bit of preflight calculation can be well worth the time, using the information gleaned from the winds aloft forecasts. In general, the winds aloft turn more westerly and increase in strength as altitude increases, making it natural to fly eastbound legs high and westbound trips low. But there are a lot of tricks up Mother Nature's sleeve, and you will find days when normal rules don't apply. Few trips are exactly upwind or downwind, so it's important to do some calculations to see how much effect you're going to get at various altitudes. Learn to choose the flight level that helps most or hinders least.

LEARNING TO LEAN

Oddly enough, **there are still pilots flying who use the mixture control only for killing the engine after parking.** Perhaps if it were colored money-green and had a dollar sign painted on it they'd take more notice. We all know the primary reason for having a mixture control in airplanes is to provide the best fuel/air ratio at higher altitudes. Yet it's a plain fact that fuel runs through the pipes slower as the knob is pulled further out, and because gas costs money, you can save as long as you lean without damaging the engine. The surest way to lean the mixture is by investing a few hundred

Even the efficiency of the Cessna 150 can be improved upon with attention paid to the information in the POH and the mixture control.

dollars in an exhaust gas temperature gauge (EGT). In less than a year the device will pay for itself, and maintenance is usually limited to occasional replacement of the probe in the exhaust stack.

Leaning with the EGT gives a reference point, or peak temperature of the exhaust, so you know when you're getting too lean for the engine's health. While some engines may be run at the peak EGT at low cruise, even leaning to 100° on the rich side of peak still saves plenty. Without an EGT, you can only lean cautiously until engine roughness first appears, then enrich the mixture until the engine runs smoothly again. Most likely, you'll then shove the mixture forward another quarter inch or so to keep the valves cool, and vow once again to buy an EGT so you'll know what you're doing.

A DIRECT ROUTE

Seeking out the airplane's best mpg setting, choosing favorable cruise altitude, and learning to lean are just starters. There are many other money-saving measures available to the cross-country pilot. When possible, fly a direct route with no doglegs. A simple GPS receiver can help generate a direct path to your destination. The poor man's RNAV (area navigation) is still a pencil line drawn with a yardstick, and it works just fine. A 10-mile saving here and there adds up to a free trip every so often, so make an effort to straighten out your cross-countries. If you stick to that straight line, you'll be dollars ahead. Just be sure to avoid restricted airspace.

CLIMB, CRUISE, AND LETDOWN

An efficient departure requires that you become established in an on-course climb as soon as possible. Request the preferred direction from the control tower as you report ready for takeoff, not after liftoff. At noncontrolled airports you should conform to the normal flow of traffic, but you can save time by making your departure from a tight traffic pattern rather than a leisurely meander over a 5-mile circle. If you're departing in a direction opposite that of takeoff, make the turn back to the downwind leg as if you were remaining in the pattern, staying within a mile or so of the field. Then simply exit the pattern by climbing out the top and make any changes in heading after safely clearing the local traffic.

Because your best efficiency lies at high altitudes, it would pay to climb as rapidly as possible to the cruising level. Unfortunately, best rate-of-climb speeds tend to produce nose-high pitch attitudes that are uncomfortable for the passengers and block forward visibility. It's wise to seek a deck angle that lets you keep the horizon in view, even if it takes an extra minute or two to reach cruising altitude.

Airplanes with fixed-pitch propellers may suffer very little by climbing at an airspeed greater than V_y, because the increased rpm (revolutions per minute) of a higher airspeed generates more horsepower, which translates into more climb. Full throttle is commonly used during climb in these aircraft, because the propeller pitch chosen by the manufacturer is usually best suited to cruise rather than climb, which limits the rpm available at low speeds. Airplanes with constant-speed props are more often than not climbed at reduced power settings to reduce the noise and wear of full-power operation. Some leaning may also be permitted in a reduced-power climb. However, climb power should be as high as practical for the most efficient climb to altitude; a minimum of 75 percent power would be recommended. Remember to advance the throttle to maintain desired manifold pressure as altitude increases until reaching the limit of throttle travel—usually 5,000 feet or so at climb power settings.

Unfortunately, ideal cruising altitudes are not always available because of weather or the limitations imposed by air traffic control if flying IFR. Any ice accumulated by climbing into a cloud deck can nullify the gain in efficiency—another reason to stay VFR. On the other hand, smooth air can provide 5 to 10 mph more true airspeed at the same power settings, making a climb to get on top of the rough stuff quite advantageous.

As you see your first checkpoint come into view, double-check your exact relative position. When fuel was cheap, it mattered little if one got a few miles off course now and then. It costs a lot more to get back on course at today's prices, so try to stop those deviations before they exceed a half mile. If you're south of that little town when you're supposed to be north of it, make a slight heading change to return to the pencil line now, not farther along in the flight after you're off course 5 miles downwind. When flying by the VOR, back up the CDI needle with your plotted courseline; at long distances from the station there can be considerable difference in where your VOR needle says you should be and where you want to be. Fly the courseline, not the needle. If using GPS guidance, match the bearing and track figures to assure a straight path to your destination.

Now do some planning for the arrival before you actually get to your destination. It takes several minutes to let down from high altitudes, particularly if your passengers are new to light airplanes and must be descended at sedate rates. Plan a 500-fpm (feet per minute) descent at cruising airspeed and project where you should begin letting down to arrive at traffic pattern altitude just as you reach the airport. You might be surprised to learn that a descent from 10,000 feet may need to begin 30 miles out. No doubt the greatest efficiency would be realized by shutting the engine down at cruise altitude and deadsticking the last few miles, but that's a bit extreme, even at today's cost of operation.

The best method for perfecting your cross-country technique is to experiment on a route you fly regularly, keeping a log of your fuel consumption over the various trips. You may soon discover there's free flying available if you're willing to use your head.

THE FLYING
VACATION

You deserve it. You've been working like a Trojan for months. You're sick of getting up in the morning to face the same old rat race, and besides, the spouse and kids want a vacation somewhere, too. This is as good a time as any to take off and get away from it all, to forget the mess at home and soak up new sights in unfamiliar surroundings. And **since you're a pilot with a magic carpet at your disposal, why not make it a flying getaway,** just like in the sales brochures?

WHAT TO TAKE ALONG

Before you start fitting a station wagon load of equipment into the airplane, some aerial vacation planning might be in order. Right away you must face the hurdles of space and weight limitations; you must reassess your baggage load in terms of the airplane's capability. You have probably been flying most of your cross-country flights with only a change of clothes and a toothbrush. This time you may be gone for weeks, and you must choose wisely to make maximum use of the available room.

Naturally, if you're used to traveling by any method, you will know enough to pick easy-care clothes, wash-and-wear items that will stretch your suitcase capacity. A winter vacation will require extra weight and space allowance for the bulky coats and sweaters of the season. Summer flying, on the other hand, can mean doing battle with high-density altitudes, another valid reason for careful weight and balance planning. If you're flying a four-place airplane, figure on using only three adult seats, or limiting the fuel load, to permit hauling vacation bags. Also, vacationing airplanes have a way of coming home with extra goodies stashed aboard, so don't load right up to gross when shoving off.

Most likely you will be required to set the date of your departure some time in advance, vacation schedules being what they are. Because you

Packing for a long vacation trip requires planning and a careful examination of weight and balance.

don't know exactly what the weather situation may be months or weeks ahead, have several routes to pick from. In case you can't get to Reno, try for Las Vegas. Anyplace is better than sitting around home waiting for the destination weather to clear. This is the secret of enjoying a flying vacation: Maximize the airplane's advantage—a rapid change of scenery—without locking yourself into a rigid schedule.

An evening departure after work can be a perfect way to start a flying vacation, depending on the weather pattern. It should be possible to fly several hundred miles in pleasant flying conditions before the evening gets too far advanced. Stop short of extensive night flying, especially in unfamiliar country. But, with daylight savings time spread over much of the year, that still leaves 2 or 3 hours for a getaway in broad daylight.

Accumulate the trip's baggage well in advance of zero hour, perhaps loading the airplane partially the night before, so something doesn't get overlooked at the last minute. Separate occasionally needed items, such as a hair dryer or travel iron, from RON essentials like shaving kits or makeup bags. Normally, if you pack correctly, only one bag per person will need to be hauled to and from the motel at overnight stops. Place inflight requirements, like munchable foods and games for the kids, in a separate box for the cabin. Keep cameras, batteries, and film or storage media in their own bag, so you won't miss a shot because they were inadvertently buried in the baggage hold.

Carry some emergency gear on board, such as a set of jumper cables to help get going again after you leave the master switch on, as well as a spare airplane key wired somewhere outside the cabin. Whether or not to carry a full-blown survival kit will depend somewhat on the terrain you are to cover and the time of year, but it's always a good idea to be ready to walk out of anything you intend to fly over. Drinking water is the most-overlooked item in survival gear; square plastic jugs are light and pack easily.

A cheap set of air mattresses and blankets, or even a light tent, are good items to have along if space permits, just in case you are forced to RON with no accommodations available. The floor of the pilot's lounge becomes much less foreboding with an inflated mattress and a warm blanket. A can of insect repellent can also make life in the wild more bearable, even if you hadn't planned on living next to nature on your trip.

Emergency aerial latrine facilities are better to have and not use than vice versa. Handy commercially available containers can be had from pilot supply houses, or you can simply toss in an empty milk carton or a coffee can with a snap-on lid. Anything will look good in a pinch; just don't pretend the need for such items will never come up.

Chewing gum and earplugs make traveling much more pleasant, and some Dramamine or other remedy should be kept on hand for the *mal du air* that an afternoon ride in the back seat can induce, along with sick-sacks for the ultimate bad scene. Keep everyone busy and entertained and the time will fly by uneventfully.

Take extra cash or traveler's checks along, as well as the basic credit cards to access an ATM, because you may run into things such as a tire that needs replacement, and you may not have the desired credit card.

SHOVING OFF

When engaging in that enjoyable pastime of pretrip planning, don't overestimate your passengers' flying endurance; it's a rare crew that can withstand a 3-hour nonstop flight. Make your hops an hour and a half or so in length, stopping for a stretch even if the airplane is still amply fueled. Everyone will enjoy the trip more, and you'd be surprised at the interesting stops that just happen to be found about 200 miles apart.

For the best riding conditions, try to make the long flights during the morning hours, before the midday thermals and surface winds start kicking up turbulence. Sightsee during the remainder of the day, and perhaps make a short late-afternoon flight to the next stop, after the vertical air movement has subsided. Don't plan on flying all day to keep a rigid schedule, unless everybody is a seasoned traveler. Not only will you have to endure some turbulence to keep up the pace, you'll be likely to encounter some vastly different weather systems on a long day's journey and you may not be able to reach the planned destination.

KEEP IT FLEXIBLE

By flying only half days you will keep the schedule flexible, and you can always switch to afternoon flying to make up for a fogged-in morning without disrupting the entire trip. Naturally, leave several extra days unplanned for the inevitable weather delays or perhaps for lingering at a scenic spot that's just too beautiful to leave. Insofar as possible, make no firm dates in advance; you call for motel space when you get into town, and if everything is booked up, fly 50 miles down the road to RON. A fully planned itinerary, on the other hand, will mean countless long-distance phone calls to reshuffle the schedule when delays occur. As cheap insurance, file a flight plan while gallivanting around the country on vacation.

Don't figure on making every stop you have planned. Always lay out more than you expect to cover, so you'll have plenty of other places to go when one spot turns out to be dull. An airplane is a magic carpet, waiting to whisk you to a different world only a few hours away, without all the tedium of endless hours on the turnpike or in airline terminals. There's something to be found everywhere, and if you are forced to stop at an unplanned airport, ask around for the local attractions.

To keep ground transportation costs to a minimum, try to find a motel with airport pickup service near the scene of the first day's action, and use a rental car for the following day, rather than keeping it parked idly. See if public transportation or courtesy cars are available before dialing up a cab. Flying vacationers should wear walking shoes to make up for their wheelless state when grounded.

When negotiating new country, ask local aviator types about preferred routes and areas to avoid. Maybe a pass that looks good on the chart isn't too easy to negotiate when seen firsthand, or perhaps one lakeshore will offer better landing facilities than the other. If weather seems to behave differently than it does back home, get the opinion of the old-timer loafing in front of the line shack. It doesn't hurt to have all the help you can get, and you can reciprocate by helping strangers at your own airport.

Lightplane vacationing is easy, safe, and fun when properly undertaken. Don't expect your first trip to be like the same excursion in an automobile or on an airliner; just accept the benefits of general aviation and plan around its limitations. You'll have a grand time that will keep you fired up for next year.

LOW-LEVEL FLYING

Pilots generally equate altitude with safety, and rightly so. Altitude is like money in the bank that can be drawn upon for such things as regaining lost airspeed, pulling out of a goofed-up aerobatics maneuver, or increasing gliding range in the event of a forced landing. Even in smooth air, my usual minimum is 1,500 feet above the terrain—high enough to clear most obstructions and adequate for crossing small towns without a detour. This would let me reach an emergency landing spot up to 2 miles away, or 1 mile with enough altitude for a traffic pattern.

WHEN YOU'RE FORCED TO FLY LOW

However, **there are times when you just can't fly as high as you would like** and must sacrifice some of your altitude if you are to continue the flight. A VFR-only pilot must stay out of the lowering clouds along the route, as an example. Quite often a precipitation ceiling with light rain will reduce in-flight visibility to less than a mile when flying near the cloud base. Dropping down 500 feet or so can restore visibility to several miles— enough to find an airport and land. Clear weather with strong winds aloft is another challenge; it's not uncommon to find headwinds of 40 to 50 knots 3,000 feet above the surface and only half that near the ground. The only way to maintain a reasonable groundspeed is to seek the lighter winds down low, which is perfectly safe if the terrain is not mountainous. Low-level flying does have some peculiar hazards, however, and we trust the old-timer won't laugh too much if we give some of his or her tips on what was once considered everyday flying technique.

HOW LOW CAN YOU GO

There are limits on how low you can fly. The Federal Aviation Regulations (FARs) say to maintain sufficient altitude to negotiate a forced landing with-out undue hazard to persons or property on the surface, a "sufficient" amount

There are times when flying low is the only way to go. At the legal minimum of 500 feet AGL, one must take care to avoid houses such as this one hidden in the trees, out of consideration for the noise impact on residents.

evidently being left to the pilot's judgment. The regulations do require at least 500 feet above ground level outside congested areas, which is plenty low enough. I prefer to hold my low-level emergency flying to 800 feet in deference to any airport traffic patterns I might encounter. Detour airports, whatever their size, rather than bore straight through the traffic pattern at low altitude. Using a normal traffic pattern altitude also has benefits if a forced landing becomes necessary, because the approach legs will begin at a normal height. Keep a watch for rising terrain elevations on the chart, and update your altimeter setting whenever possible; you've little margin for error here. If you're sneaking along under lowering weather, set yourself some definite altitude minimums for a 180° turn. If you are forced to drop down from 1,000 feet above ground level (AGL) to 800 feet or so, start planning a course of action right then, and when pushed down any further, get out without hesitation. The ceiling may be coming right on down, and you need time to find an airport. If you wait until 500 feet becomes 400, then 300, it may be too late to do anything other than risk a landing in the first pasture you can find.

NAVIGATION

The most noticeable change from everyday flying will be the navigation techniques. Your range of vision is much shorter at low altitude, and checkpoints are difficult to spot until they suddenly appear under the wings. A highway intersection, for instance, lies hidden behind bordering trees or hillsides, and if you miss it by a mile you may never see it. At low levels the apparent ground speed is much faster, which means a checkpoint will be gone from view in a fraction of the time usually available for identification. Don't be afraid to circle the landmark in doubt for closer study; note the heading you were following before you started the turn, however.

Rather than blundering along half lost, you'll soon find it becomes much easier to follow a well-marked highway or railroad that happens to be going in your general direction. Never mind the jokes about the "concrete compass" or the "iron beam"; at least you'll know where you are. I do suggest following such a route just to right of center—both for a good view of the roadway and to minimize chances of a head-on collision with someone following it in the other direction.

In areas where there isn't anything on the ground to follow, one falls back on dead reckoning by clock and compass. To avoid getting hopelessly off course, note prominent landmarks—such as towns, lakes, roads, or rivers—that lie along the route a few miles from the courseline. Then, if you find yourself over one of these points, it will be readily evident that you need to alter your heading to get back to the direct course. This is, quite simply, bracketing yourself so you catch the error before it gets out of hand.

What happened to the radio all this time? As you no doubt recall, the very high frequencies (VHF) are subject to line-of-sight limitations, and the signals can be blocked by high ground and curvature of the Earth at low altitudes. Over flat country you can get a course needle indication for 50 or 60 miles from the station, but beware if the to-from flag is not pulled into view. Such weak signals are not always accurate and should not be relied upon for precise determination of position. The automatic direction finder (ADF), bless it, is still usable even at ground level, as long as thunderstorms don't create false bearings, and the GPS receiver works as long as there are birds in view. In the absence of these last two aids, you are on your own when cruising down on the deck, where you cannot rely on electronic guidance from the VOR or communicate reliably.

DANGERS TO AVOID

Now there is the matter of obstructions. Low-level flying would be fun if it weren't for the continuing proliferation of nearly invisible towers across the land. They pop up without warning as you fly along in the middle of the countryside—lighted and painted to be sure, but still mighty hard to see. Don't expect the sectional chart to keep you out of trouble. It may take a year for a tower to make it onto the chart, even if you fly with up-to-date charts.

I had to learn about towers the hard way, so listen and learn. I was hired to deliver a man to a lakeside airport early one fall morning. We left at sunup, under ceiling and visibility unlimited (CAVU) skies, but as we neared the lake country a solid blanket of fog appeared. This wasn't unusual for the area, so I simply continued on a dead-reckoning course above the white blanket, watching for a hole in the vicinity of the airport. Alas, no holes appeared, and I wandered around for a few minutes, trying to give the appearance of doing all I could to get my passenger to his destination before giving up. I had already told him we were going back home when I spotted a lone hole with a certain marina's name painted on a boat dock roof. Instantly I was oriented; the airport was only a few miles down the road from the marina.

We spiraled down through the hole, finding a few hundred feet of murky airspace underneath the lifting stratus, and roared off down the highway to the airport. Sighting a windsock above the trees, I pulled off the power, coasted around a low, tight traffic pattern, and deplaned my shaken passenger after what I thought was a superb job of flying.

A local pilot strolled up and said, "Pretty low, isn't it? Watcha do, make an ADF letdown on the radio station?" He glanced down the highway in the direction I had come, and there stood a splendid example of the steelworker's art, poking up into the ceiling. Had I chosen to follow the road along the north side instead of the south side, I would have hit that tower dead-on. It was there on the chart, but who looks at charts when watching

Down on the deck, because of weather or strong headwinds aloft, this Citabria pilot sticks to open country and watches for obstructions.

for an airport? I mumbled something about ADF to the man and slunk away, suddenly much older and wiser.

Therefore, plan a low-level flight as if you were heading into mountainous terrain, checking all along the route for maximum elevations to be cleared before choosing an altitude for flight. Don't let your eyes leave the windshield for more than a second or two, just in case an uncharted new obstruction looms into view. Fast airplanes are particularly hard to manage down on the deck, and it is only good sense to back off a little from max cruise in these birds, to give yourself more time to dodge.

When flying in canyons or valleys, be alert for electrical transmission lines strung across from rim to rim. Some states are requiring these lines to be marked with lights and orange globes, but most are noticeable only by their supporting towers. Don't try to go under the wires, judging height by the towers or poles; a long span always sways very low at the center, making successful attempts pure luck. And just in case you've forgot, remember the cardinal rule of terrain flying: Never fly *up* a narrow canyon, always *down,* toward lower ground. The merging forks of stream beds will point the way downhill, if in doubt.

As a participant in many Civil Air Patrol search missions, I can testify that low-altitude flying isn't for kicks; it's a hard, demanding job. Use it wisely, however, and learn its pitfalls, and it might bring you home someday.

DODGING TALL TOWERS

As the sun's light scatters in the dense haze, the pilot peers into the milkiness through a windshield long since rendered semiopaque by impacted insects. The visibility is reported to be 3 miles at the surface, but at an altitude of 1,000 feet AGL it's anybody's guess. He checks the sweat-stained chart on his knees for the hundredth time, trying to match the chart's symbols with a village or turnpike in the circle of restricted visibility beneath the aircraft.

He isn't ready to consider himself lost, of course; few people will be honest enough to admit such a plight until it's too late for effective action. But it's been quite a while since he's felt that satisfied glow that comes with a solid fix. The GPS says he's 74 miles away from the destination, but that doesn't tell him where he is at this instant.

Glancing up, the pilot catches the blur of a rapidly approaching image, all too close, at 12 o'clock level; it's *a tower,* a tall spike of slender steel suspended by a far-flung spider web of cables—and it's uncomfortably near. He cranks in aileron and elevator to escape the intruder's clutches and the top of the structure slides by off the right wing, a hundred yards away but still a near brush with tragedy. Shaken, the pilot climbs an extra thousand feet to guard against encroachment on his airspace, but in doing so sacrifices still more ground contact to the all-encompassing haze.

We've all been that pilot at some time or another—a little off course, in poor visibility conditions, and a tad too low in an attempt to orient ourselves. Sometimes the pilot isn't so lucky and the tower snares the unwary aviator. It isn't just haze that can bring us down into the reach of obstructions; a low stratus deck over rising terrain can tempt a "gotta get there" individual into ducking under. Freezing temperatures at higher altitudes can dictate a low-level route, or the pilot may be letting down for a landing while still a few miles from the airport so he'll have time to slow down for the pattern.

PREFLIGHT PRECAUTIONS

Whatever the reason, **recognize the potential for disaster when you enter a thicket of steel near the ground** and take steps to minimize the danger. It would be easy to say you'll never fly lower than 1,000 feet AGL, the breakpoint between large and small obstruction symbols on the sectional chart. However, that's no guide to safe operation in itself. A small symbol can mean a 999-foot tower and the tall-tower symbol only warns of an obstruction *taller* than 1,000 feet; it doesn't say it can't be 2,000 feet tall, as in the case of the towers just off the airway east of Springfield, Missouri. Also, it's easy to overlook an elevated bit of terrain decorated by a relatively small 500-foot tower, but that combination of hilltop and steel spike are enough to bring the top of the tower into frequently used airspace. Be sure to judge terrain elevation by the ridges, not the valleys.

Of course, sectional charts have large numbers in the middle of each lat-long grid, signifying an altitude increment that will clear the highest known object in that block by 100 feet or more. A "12," for instance, calls for a minimum flight altitude of 1,200 feet mean sea level (MSL), which could mean an obstruction as tall as 1,099 feet MSL could be found in that particular grid. Because altimeters can be as much as 70 feet out of tolerance, this leaves scant room for error.

The tower just ahead of the wingtip measures 1,444 feet AGL, but it isn't the only one around. Barely visible in the center of the picture, right of the curve in the road, is a 1,626-footer some 5 miles away, and two smaller towers may be seen near the bottom center of the picture.

For protection without constant checking, the sectional's title block carries a warning of the highest-known terrain or obstruction on the chart, but this advice is largely wasted in western states, because one side of the map may depict terrain 10,000 feet higher than the other side, where the warning would have little significance. The Jeppesen or NOS low-altitude en route charts also provide some obstruction-height warnings, as MEAs (minimum en route altitudes) on the airway structure will give an obstruction clearance of at least 1,000 feet along the airway. Approach plates bear one or more MSAs (minimum safe altitudes) for the maneuvering area, generally valid for a 25-mile radius of the final approach fix.

KEEPING WATCH IN FLIGHT

Well-known metropolitan towers might seem to present less hazard than the countryside's crop, but in reality any safety advantage will probably be negated by the thicker traffic density around the city antenna farm, even though its location is very well known. However, one simply doesn't expect to encounter a 1,200-foot needle out in the vacant spaces, where the pastoral scene lulls pilots into lethargy. The guy wires will normally reach out 80 percent of the tower's height from the base, so a 2,000-footer will have wires spanning nearly five-eighths of a mile, even more reason to give tall towers a wide berth wherever they are encountered.

Free-standing microwave towers are more easily spotted than guyed towers, because of their heavier structure; they tend to be shorter as well, although there is a 1,048-foot free-standing TV tower just off the end of runway 19 at Kansas City's Downtown Airport. Haze can make even thin towers stand out in relief against a white background, as opposed to the ground clutter that camouflages them on a clear day. Even so, the snow-blind pilot flying in haze might not see the steady-state approach of a tower with her name on it. It's the ones that don't change relative position that get you, so take a peek behind the windshield post periodically.

You generally should not fly at low altitudes at night, but towers are easier to spot in the open countryside during hours of darkness. However, they are sometimes camouflaged in well-lit city areas. Dawn or dusk is the worst time for spotting landmarks, of course, towers included, particularly if one is looking into the sun.

Placement of towers is certainly not done without restriction; proposals for construction near airports must meet clearance-slope guidelines issued by the Federal Aviation Administration (FAA), which has final authority over structures higher than 199 feet AGL in the vicinity. Away from airports, the FAA has no authority to deny construction. Towers shorter than 200 feet need not be marked or lighted. If you see a notice of proposed construction posted on the local airport bulletin board, check it out in relation to your area's arrival and departure patterns; you may want to comment on the

Near a busy VOR and airway, here are three 2,000-foot towers, plus another pair measuring a mere 1,627 feet tall! The chart has a "37" on this grid block, calling for a minimum of 3,700 feet of altitude in an area of 1,500-foot terrain elevations.

proposal in order to shift the tower to a less critical spot. It's much harder to get a tower decommissioned or moved once it's built.

SUMMATION

Include a check for obstruction clearance in your preflight planning, noting safe altitudes on your flight log in advance of takeoff. Keep a sectional chart handy during all off-airways flights. Don't fly with en route charts only. Don't take it for granted that everything is charted; remember there's a lag of 6 months or longer between commissioning and updating. Few obstructions will protrude more than 2,500 feet above the mean terrain, so stick to higher altitudes whenever possible. If forced to fly right down on the deck, plan on foregoing idle conversation with passengers and devote your full attention to the approaching scenery. There's no way an airplane is going to win an encounter with a steel tower, so plan on never challenging one.

THE VFR FLIGHT PLAN

Remember the VFR flight plan, that hoary anachronism from student training days? Have you called one in to flight service lately? Probably not, because most pilots seldom think about the chances of being forced down short of their destination. The fact is, **there's still a place for the VFR flight plan** as a free trip-insurance policy that pays off big if your rubber band breaks somewhere en route.

Flying cross-country used to be a bit more challenging than it is today. Flight plans were quite popular, because if trouble cropped up, it was just a matter of time before someone would come looking for you, and the flight plan told them when and where to start. Airplanes were slower, good airports were less plentiful, and lightplanes often were not equipped with radios, so once airborne, there was no way to check one's bearings or get a weather report. The flight plan was vital for added peace of mind.

TIMES HAVE CHANGED

With improved equipment, times have changed. Nearly all aircraft have a radio or two, and we generally have good in-flight communications during the entire course of the flight. Most aircraft are required to carry an ELT (emergency locator transmitter) to broadcast a signal if there is a crash. We may even be able to get a friendly ATC controller to give us flight-following service by radar. As the term "flight-following" was used originally, it referred to a type of VFR flight plan that assumed you would check in at each VOR along your route, and flight service station (FSS) personnel would consider you overdue if you missed your estimate for any of them. Talking to center was reserved for the big boys. A vestige of the old FSS flight-following system remains as the lake-reporting and mountain-reporting services available in some areas.

Most pilots assume that in the event of an emergency landing they'll have ample time to get a call off to someone to tell them of their plight. Unfortunately, this isn't always the case. If you're fighting a weather problem, you may be forced down gradually by lowering ceilings; by the time you desperately want to let the rest of the world know your location, you're too low to communicate. Even if you have plenty of altitude, it's easy to let time and altitude dwindle away while you troubleshoot the problem or hunt for a field. As an example, I flew a search for an airplane reported to be down in the local area, and it turned out that the pilot actually had bounced to a stop on the ground before he found time to talk to FSS. Fortunately, he was only a few miles from a VOR, so he still had communications at ground level.

WHAT HAPPENS IF YOU DON'T ARRIVE?

If a problem arises and you need someone to start looking for you, don't count on getting off a message at the last minute. The ancient and honorable VFR flight plan can be a lifesaver because of its passive ability to trigger a response, and you'll save a lot of wasted man-hours in speculative search efforts by revealing your plans. If you don't arrive at your filed destination by 30 minutes after your ETA, a communications search will be started. If this is fruitless, an alert notice is sent out on the landline network to all FSS facilities along your route. If another hour goes by with no word from you, a full-blown search mission will be mounted, coordinated by the Air Force Rescue Coordination Center. The hard work of the actual search is usually delegated to the volunteers of the Civil Air Patrol, with assistance from the military, law-enforcement personnel, and local and state rescue units. The success of a search mission varies, depending on how closely you kept to your planned route and how cleverly you managed to conceal your downed bird. If your ELT works as advertised, there's a good chance that you won't have to wait until spring to be rescued.

WHEN SHOULD YOU FILE?

As an example of a good use of a VFR flight plan, let's use the old vacation scenario. Jack and Judy load up and head west, telling their coworkers they're going to California and Oregon on their vacation. One day into the trip, they meet with disaster and slide to an ignominious halt on a snow-covered hillside, uninjured but miles from anywhere. Now, who's going to look for them? Maybe after a couple of weeks roll by somebody will happen to remember the happy vacationers. "What ever happened to Jack and Judy? Did anyone get a card from them?" If Jack and Judy are incapacitated by injuries, or if the weather is extreme, their bones might not be found until the snow melts. Their ELT may go off and bring the cavalry riding to the rescue—if they put batteries in it, or if the antenna didn't tear away.

Personally, I would like someone to know my intentions, in addition to carrying the little box in the back.

But file a flight plan on every trip? Obviously, no. I doubt that the FSS network would be able to handle the load. So long as *someone* is expecting you to be *somewhere* at a certain *time*, there's little need for a more official flight plan. If you tell your spouse you'll be heading to Missoula today and stopping by Butte tomorrow, and you make a practice of calling if you're late, that's a sort of flight plan. Or if you have appointments to keep in Hot Springs this morning, your secretary probably will start getting inquiries if you don't show up. If the airplane is a rented one, there was probably some discussion of your intentions. In all cases, make sure somebody at your home base knows who to contact if you're overdue; the nearest FSS will get things started, if given the word. But whenever you're going to be out of contact with the rest of the world for an extended period of time, file a flight plan.

There will be times when filing is impractical, as when arriving at or departing from a remote boondocks airport lacking even telephones. You may not know what your ability to close the flight plan will be and you may want to stay off the record to prevent a needless search from being started. Or your schedule and route may not be clear at the time of your departure. These are valid reasons to fly without a flight plan, but even so, you can check in over a VOR now and then with a position report, or make a weather check by giving your location and intentions; this will put you on record as having been somewhere at a certain time, a fact that can help the searchers immensely.

HOW TO FILE

Filing a flight plan can be done in person at the FSS (rarely), by phone when checking the weather, on the computer if you're using a weather briefing service, or by radio once you're airborne. The latter ties up the airwaves and the FSS air-ground specialist, so it should be avoided as much as possible. Get familiar with the flight plan form so you can rattle off the required items in the proper sequence; this helps the FSS specialist write it down the first time and avoids requests for a repeat. Don't give the specialist long descriptions of what each item is; he or she knows the order of the entries, so only short guide words will be necessary.

In most cases, the FSS will hold your flight plan until you report the takeoff by radio, but if an airborne check-in is not possible (as when flying a no-radio airplane), you can be assumed off at a certain time, with an agreement to call back by phone if you can't get into the air. Once activated, the flight plan goes to the FSS responsible for your destination airport, and it now becomes their responsibility to call for the bloodhounds if you don't show up. Therefore, if you change your mind or run late, the proper party

Form Approved: OMB No. 04-R0072

DEPARTMENT OF TRANSPORTATION FEDERAL AVIATION ADMINISTRATION **FLIGHT PLAN**	CIVIL AIRCRAFT PILOTS. FAR Part 91 requires you file an IFR flight plan to operate under instrument flight rules in controlled airspace. Failure to file could result in a civil penalty not to exceed $1,000 for each violation (Section 901 of the Federal Aviation Act of 1958, as amended). Filing of a VFR flight plan is recommended as a good operating practice. See also Part 99 for requirements concerning DVFR flight plans.						

1 TYPE VFR IFR DVFR	2 AIRCRAFT IDENTIFICATION	3 AIRCRAFT TYPE/ SPECIAL EQUIPMENT	4 TRUE AIRSPEED KTS	5 DEPARTURE POINT	6 DEPARTURE TIME PROPOSED (Z) ACTUAL (Z)	7 CRUISING ALTITUDE

8 ROUTE OF FLIGHT

9 DESTINATION (Name of airport and city)	10 EST. TIME ENROUTE HOURS MINUTES	11 REMARKS

12 FUEL ON BOARD HOURS MINUTES	13 ALTERNATE AIRPORT(S)	14 PILOT'S NAME, ADDRESS & TELEPHONE NUMBER & AIRCRAFT HOME BASE	15 NUMBER ABOARD

16 COLOR OF AIRCRAFT	**CLOSE VFR FLIGHT PLAN WITH_____FSS ON ARRIVAL**

FAA Form 7233-1 (5-77)

This time-honored flight plan form still serves a useful purpose. Block 17 now requests a destination phone number to simplify searches.

to inform is not the FSS with which you filed the flight plan, but the destination FSS. If you are forced to stop midway in your flight, any FSS can send a cancellation or an extension of an arrival time to the FSS holding your flight plan.

Please make sure you close the flight plan. Don't assume the control tower does it (they don't and won't), and don't tell anyone else to do it for you. Just call your cancellation in by radio or get to a phone promptly after landing. Failure to do so may result in your being awakened from a sound sleep by a posse of very irate rescue personnel who won't take kindly to your being snug in your warm little bed when they are out beating the bushes for you.

Don't forget about your old friend, the VFR flight plan. It's still a good idea, even though an attempt to discontinue it has been made more than once. Use it whenever no one else is going to be looking for you, for your family's sake if not for your own.

USING GPS IN THE COCKPIT

W hen GPS burst on the scene late in the twentieth century, it revolutionized not only aviation, but nearly all forms of transportation and industrial tasks requiring position information. What GPS offers is a three-dimensional worldwide grid that can be supplemented with data specific to user needs. For automotive use, street names and predictive computation at low speeds are employed. For marine applications, information about buoys, bottoms, and beacons is overlaid on the base map. Hikers need terrain features and waypoint storage to blaze a trail leading back to camp. Aviation, on the other hand, requires speedy computing, fail-safe redundancy, and signal integrity monitoring, coupled to a database of existing waypoints left over from our earlier systems.

As aviation GPS has grown, it has reached into every corner of the market; portable, battery-powered receivers can put a moving map into ultralight vehicles and sport planes for a few hundred dollars, while integrated GPS engines can support all-glass jet cockpits costing hundreds of thousands. In-between, we can choose color map units that can be clamped to a control yoke and display a course to any point in the database in full view of the pilot, or panel-mounted GPS receivers that steer an autopilot, tune the appropriate communication frequencies, and let us fly instrument approaches that didn't exist before GPS. The choices are almost limitless.

Now that GPS navigation has become universal, pilots with limited experience are encountering it very early in their careers. You may have used GPS as a student pilot, and you certainly will have it in your cockpit as soon as you're out on your own. It behooves us to give some very basic explanation of these magic boxes, without being specific about any manufacturer's gear. Unlike VOR/ILS (instrument landing system) navigation, the highly complex GPS receivers have very little commonality and are capable of doing so much more than any one pilot is likely to need that some reference to the owner's manual and a few hours of practice time are mandatory.

DON'T BE INTIMIDATED

Very few pilots will use the more subtle features of even the simpler GPS receivers. What we're about here is to describe what a new user will find and how to get basic utility from the box. To recap what makes the system work, the American taxpayer has paid for a constellation of 24 orbiting satellites (21 operational birds with 3 backup spares), any three or four of which can fix a receiver's position. The satellites are not parked in fixed locations, but are moving around the Earth in a low Earth orbit at all times. Accurate timing signals and unique identifying codes allow receivers to work navigation solutions almost instantly as long as the birds are in view of the antenna. GPS won't work in a hangar or on an internal antenna that doesn't have a view of the sky, and, because of the extremely weak signal, may suffer outages during periods of intense solar flare activity.

What makes aviation GPS so valuable is, quite simply, the map presentation and the aviation database. Those two features turn navigation into a matter of exclaiming, "Look, there we are and over here is home." Once a destination is defined, you'll have a courseline, distance to go, speed and steering information, and probably en route features and airport information. The sectional chart is more and more being relegated to a backup role. Have a care, my friends; don't fail to have paper navigation handy, because

The Garmin GNS530 is among the most capable of global positioning system receivers, but even it has a simple direct to function key on the middle right side of its bezel.

someday you'll stare at a blank screen and wish you knew where you were. As with any navigation method, you should always operate with two unrelated systems checking each other—and that doesn't mean two GPS receivers.

WHAT GPS DOES

At turn on, GPS units go through a self-test initialization, often requiring you to acknowledge disclaimer messages and database age by prompting you to press "Enter" or "OK" buttons, after which the unit begins searching for satellite signals. The time for usable navigation information to appear may be several minutes, particularly in those instances where the unit was turned on far from where it was when it was shut down.

From the pilot's standpoint, all GPS receivers share certain characteristics, even if they look and operate differently. Your current position on the surface of the Earth is known, in most cases down to a matter of a few feet, shown as the latitude and longitude, or in bearing/distance reference to the last known location. There may be a little airplane symbol on a map of varying complexity, depending on the database in the unit. If asked to, the receiver can show you the way to a destination (waypoint) or a string of waypoints leading you to your landing point. This process may begin with pushing a "direct to" key, designated by a large letter D with an arrow through it, or pressing a database key or route key. Some units require activation of a blinking cursor function, after which you can dial or key in the identifier of your waypoint.

Unlike VORs or other Earth-based navigation systems, GPS does not provide course guidance outbound from a fixed point. Rather, the receiver always needs a designated waypoint ahead of the aircraft, toward which it can present a suggested course, distance to go, and deviation from a direct line. Of course, the airplane may be proceeding away from, or at right angles to, the GPS course, but the information is given in a "to" direction.

In its simplest form, other than just keeping track of your location, GPS navigation defines a line from your present position direct to your destination, the location of which is usually in the aviation database loaded in the GPS receiver. If the unit's display can show a map, staying on course is like drawing a picture on the screen by steering the aircraft. If you choose not to use the map, you'll probably have data fields displayed showing groundspeed, course to the waypoint, the direction of the actual track made good, and perhaps the cross-track error, or distance off the desired track. Most GPS displays can present an electronic version of a CDI (course deviation indicator) like you're used to seeing with a VOR, or perhaps be coupled to a panel-mounted CDI shared with the VOR receiver.

The best way to stay on course is to match the course number with the track's number; if the two match, you're going the right way across the ground, and that's all that counts. Heading is automatically adjusted for wind crab. Your cross-track error, if shown, will never be too great, probably less than a mile, if you match the numbers. If using an unslaved directional gyro for heading information, don't forget to reset it to the magnetic compass every 20 minutes or so to correct precession.

STAY AWARE, DON'T JUST STARE

The weakness of GPS is its ability to take you places you really don't want to go; hot military operations areas, restricted and prohibited areas, temporary flight restrictions, and sparsely settled areas. It's up to you to keep track of your location, even if your GPS has special-use airspace in its database. Not every restricted airspace can be shown, and more pop up every day. Stay aware, don't just stare.

From this basic here-to-there utility, GPS designers have added tons of features like accepting and storing multileg flight plans, popping up a list of nearest airports or nav fixes, and accessing nice-to-know things about your destination like its frequencies and runways. Don't feel obligated to use all the unit's capabilities; hardly anyone would, except the technically oriented. Beyond the pages of screen options, there are setup options that can be used to customize the receiver to your tastes. Or, you can just leave the default settings in place, as most of us do.

GPS navigation has revolutionized the way we do business in aviation, and there's no doubt it's enhanced safety by keeping pilots from getting so badly lost they run out of fuel. At the same time, it tends to encourage boldness in ways we never thought of before, like pressing on in visibility of 3 miles or less with assurance that the airport on the moving map will show up in a minute or two, or crowding closer than we should to the lateral limits of Class B airspace. As with any tool, GPS must be kept in its proper place and used with supplemental information. Don't let it replace good judgment.

TRAFFIC AND AIRPORTS

MIDAIR MENACE

There is a continuing need for us pilots to remind ourselves that other airplanes are up there with us. One of the most commonly cited reasons for flight test failure is a lack of vigilance by the applicant; other airplanes come into view and pass with no recognition because the pilot is deeply engrossed in the cockpit tasks. I'd have to call it a fair bust, even if it were one of my own students, because I might be up there myself someday, and I sure want the pilot to be looking for me.

STATISTICS

What is a midair collision, typically? Statistics show that about 85 percent of all recorded midairs occur between small general aviation aircraft in clear weather. The chance of a midair occurring while both aircraft are under IFR control in actual instrument conditions is rather remote and, despite all the headlines and movie hoopla, airliners do not run a great risk of being attacked by little airplanes. Large, high-performance airplanes are generally separated from the smaller general aviation fleet by reason of operating desires; the jets need to go high as soon as possible, whereas we normally want to remain as low as practical.

The 25 to 35 persons killed each year in some 50 midair collisions are a miniscule part of total aircraft accidents, a far less serious problem than weather accidents or poor flying skill during takeoff and landing. Any accident is one too many, of course, and it behooves us to take such steps as are reasonably proper to prevent it. Understanding the circumstances behind most midairs will help alert us to the problem, perhaps motivating us to look around a bit harder.

WHERE DO MIDAIRS OCCUR?

The large majority of collisions happen on or within a few miles of an airport. An old railroader once observed of airports, "We have one track out on the main line but several of them coming into the station. You fellas have lots of tracks up there but only one here at your depot. That ain't too smart." Unfortunately, he's right; the common target of all airplanes converging on an airport is the approach end of the duty runway, making it only natural that collision accidents will arise during this phase of operation.

Although midairs do occasionally occur within controlled airport traffic patterns, the bulk of them happen at small, noncontrolled fields, where discipline often slips and pilots tend to do their own thing. VFR daylight weather usually prevails, although we would imagine that haze contributes to many of the midairs. California has always led the country in midairs, logically enough, because it has a high aircraft population in areas noted for low-visibility conditions.

Often the two tangled airplanes were maneuvering for takeoff or landing when they got together; one may have been taking off while the other was in a nose-high approach to landing, or both airplanes may have been on final approach at the same time. I once experienced a near-miss at a noncontrolled FSS-only field, reporting on a 2-mile straight-in final only to hear another airplane call a 2-mile final a few seconds later. He and I compared notes hurriedly, and found that he was above my left wing and I was below the right side of his instrument panel, converging nicely. That was my last straight-in approach at an uncontrolled field.

Dual instruction plays a major role in midair collisions, because both occupants of the airplane are busy teaching and learning with minimal time

The overtaking aircraft, descending at high speed, was spotted just in time to avoid a midair collision. The slower airplane is banking away, apparently undetected.

left over for looking outside. Flight instructors hold the key to instilling vigilance in future pilots, however; if CFIs make a point of looking outside themselves, locking the control wheel in neutral until the student has cleared the turning airspace, and otherwise emphasizing the seriousness of the matter, an instinct to look will be fostered in the student. On the other hand, if they sit there with paralysis of the neck and eyeballs, so will their students.

HOW TO AVOID MIDAIRS

One of the best ways of avoiding near misses with high-speed IFR traffic is to fly where they aren't. Learn which runway is active when passing near a busy field, and which approach is in use, by listening to ATIS (automatic terminal information service broadcast) or the tower chatter. Monitor the approach control frequency in your area even if not in contact yourself. With an idea of where the inbound traffic is coming from, you can steer clear of the "dump areas" of descending jets. Just make sure your deviations around the big airport don't take you unknowingly through a small airport's traffic pattern at low altitude.

The perfect visibility aircraft does not exist; therefore, we must recognize that our airplanes have blind spots and clear them for traffic constantly, particularly in the vicinity of an airport. Adjust your seat for best vision, if it cranks up and down; too high and you may not be able to see out the side windows without ducking, too low and you will have a large area obscured by the panel. Raise a wingtip to check for descending traffic in your high-wing airplane; dip a wing or turn slightly to look underneath your low-wing bird. Lean across and look out the right side of your airplane if you don't have an eagle-eyed passenger sitting over there to help you. If you see a shadow on the ground that isn't yours, find out where the bogey is—*fast.*

Remember that converging airplanes do not move in relation to each other; this makes it doubly hard to spot a midair threat. First, moving targets are picked up much easier than stationary ones by our scanning glance and, second, if the target is in a blind spot (such as windshield posts), it tends to stay there. A clean windshield naturally makes such motionless targets easier to detect, but don't forget the side windows also. Extra strobes and high-visibility paint colors will help, if you're in a position to determine the airplane's outfitting. However, even the most dull-finished airplane can be seen before it hits you, if you're looking.

LOOKING AROUND

Not looking at all is the biggest midair hazard. As a CFI, I ride with a great many pilots and habitually evaluate them; very few of my flight review applicants, for instance, exhibit adequate vigilance for other aircraft.

Scanning must be methodical; clear the airspace you are about to occupy before you go there. If making a left turn, don't just glance out the left window; look back over your left shoulder as well for that airplane in front of which you're liable to be turning.

When making a landing approach, don't stare at the runway while on base leg; look up final approach once in a while. Don't just take the runway after completing your runup with a quick glance at the downwind leg; look at the final approach and base legs as well, even if it takes a full circle to do so. It takes only one unsettling experience, such as having an airplane zoom down right in front of you just after you've opened the throttle for takeoff, to convince you to look around more. If you're careless, you may never live through all the bad experiences to complete your learning!

Don't be tempted to make unauthorized right-hand or convenient straight-in approaches just because you think nobody is around; the few seconds saved aren't worth the risk. Enter patterns at the proper altitude, usually 800 feet AGL, so all other airplanes will be visible out of your windows instead of hiding above and below a wing or windowline.

Use the radio for position advisories, but do not rely on such calls exclusively; there are no-radio aircraft operating in our sky, and there is always the klutz who is still on center frequency when he hits the pattern. Look outside even while you're announcing your intentions over the wireless. Don't forget to turn on and check the anticollision lights, which are required to be operating even in daylight hours.

We can seek all possible help from ATC radar for traffic advisories, and we can equip our airplanes to the hilt with lights and radios, but the simple fact of avoiding most midair collisions is the need to use our eyeballs. There are other airplanes out there, so don't get complacent just because you don't see them. It only takes one to ruin your day!

UNCONTROLLED AIRPORTS

"**H**ey, look, here he comes again!" called the acting ramp scout at the small country airport. His function was to alert us to any inbound aircraft while we occupied chairs in the deep shade, sparing us from sunburn and sore feet. The subject in question had already made several passes at what seemed to be a perfectly ample amount of runway, wisely adding power for the waveoff when he turned out to be higher than a tobacco barn at midfield.

After the first go-around we made book on the number of attempts he would require, and so far the holder of the lucky number four had resigned himself to enriching the eventual winner. As the object of our wagers came out of his base-leg turn, however, another factor entered the equation in the form of a light twin on a long straight-in final approach. "They're going to get close," opined the ramp watcher, and we rose from our chairs to assume a sprinting position.

The twin gave no sign of noticing the aircraft on a close-in base leg, and the other pilot had all of his attention fixed on the runway of his heart's desire. He reached the final approach course slightly ahead of the twin, and as he banked onto final the twin's pilot saw the flash of movement and pulled up. Another midair was averted—one less headline to provide grist for the FAA's plans to expand positive control.

The hapless go-around veteran made an uneventful full-stop landing this time, but as he turned around on the runway to taxi back to the hangars he encountered—who else—the light twin on short final. Following his waveoff, the pilot of the twin had made a tight 360° turn on final at an altitude of 300 feet AGL, maximum, and did not count on the preceding aircraft still being a factor. The taxiing pilot headed for the grass just as the twin started his second go-around—more wasted gasoline, more heated tempers. The crowd returned to the chairs, ready to break up the fight when the twin's pilot caught up with the other aviator.

"Didn't think I'd ever get down," said the lone occupant of the first airplane as he disembarked in full view of our group. "Could I getcha to sign my logbook? I'm running a little behind on my solo cross-country with all these go-arounds." One of the locals moved quickly to oblige, anxious to finish the act before the twin pilot shut down.

"Hey, don't anybody listen to the radio around here?" bellowed the many-motor's driver as he climbed out onto his wingwalk. "I called three times and nobody answered me." He cast a withering look in the direction of the student and headed for the lounge, leaving his airplane in a position that blockaded two hangars, the student's airplane, and part of a taxiway. It was time for action and education.

"Come on, boys, let's get this kid out of here," I said. "Get on that wing and help me push this heavy iron back a few feet." I squinted toward the office, hoping the twin pilot was watching. Fortunately he left his brake off, so I did not have to soil his carpet with my dirty boots. I gave a few words of encouragement to the student about working on his approaches, and pointed him back toward the big city airport from whence he had come. Then I walked over to the lounge to look up the twin's pilot.

He had cooled off only a few degrees, still mouthing about the silence on Unicom and lame-brained country cowboys who wouldn't get off the runway. "What frequency were you on?" I asked. "Unicom 122.8," he said. "Isn't that what they use here?"

"Nope. The sectional I used to find this place had the frequency 122.9 shown as CTAF [common traffic advisory frequency], and that's Multicom for plane-to-plane advisories, so I used that channel for safety messages, according to the *Aeronautical Information Manual*."

"Guess I should have reported my straight-in approach on that channel," he said. What he *didn't* say was that he probably hadn't unfolded a sectional in some years, preferring IFR en route charts and approach plates for his information.

I bored in. "No, what you *should* have done was fly at least part of a traffic pattern. A straight-in approach is a great way to meet people...in midair." We sparred a few more rounds, finally agreeing that you certainly have to be careful around these small airports, and parted more or less friends.

EXPECT THE UNEXPECTED

With the lack of emphasis in training on defensive flying at small public-use airports, we can expect more and more scenarios like the foregoing. The training texts furnish 40-year-old illustrations of 45° traffic-pattern entry paths, segmented circles, and wind Ts, but today that's only half the story. Pilots must be prepared to cope with the nontextbook happenings that can arise every day at uncontrolled airports.

If there happens to be a couple of other airplanes in the pattern, who goes first? If a go-around becomes necessary, how is resequencing accomplished? How close does one operate to another airplane?

Located near a residential area, this convenient small airport has Unicom and two runways. The lack of taxiways means one should give way to landing aircraft when taxiing out for departure.

These are real-life questions at nontower airports. Nobody is up there in a glass cage, playing referee. That leaves the pilots to make all the decisions. Unicom is just an advisory channel, an electronic wind sock, nothing more. The first rule to remember when operating at an uncontrolled field is to take nothing for granted. Be expecting the other pilot to do something stupid, and keep an eye on him or her every second.

Do not assume that your traffic advisory calls were heard. There are still plenty of airplanes out there without radios, or if they have one, the volume may be turned down or it may be on the wrong frequency. Nevertheless, it's wise to make a call to Unicom or "Podunk traffic" before entering the pattern as well as before taking the runway for departure. The suggested calls on each leg of the pattern, clearing the runway prior to taxi, and so forth, are even better, if you have time.

WHAT IS A TRAFFIC PATTERN?

A traffic pattern is a rectangular box around the airport, a path airplanes should follow to maintain an orderly flow of traffic. This simple definition gets ignored by many pilots, unfortunately. The FARs have very little to say about traffic patterns, mentioning only that all turns shall be made to the left unless otherwise specified and that right-of-way belongs to the airplane at the lowest altitude. By convention, entries to the downwind leg, that part of the pattern flown beside the runway opposite to the direction of landing, should be made at midfield on a 45° angle, making it easier to avoid departing aircraft and to merge with traffic remaining in the pattern. Height of the traffic pattern varies with location, but in noncongested areas it is most commonly 800 feet AGL. It does help immensely if all aircraft enter at the same altitude so that high- and low-wing airplanes can see each other.

If you don't know the wind direction, take a look at the wind sock before committing yourself to land. If the sock is limp, winds are calm, so keep an eye on the opposite end of the runway for opposing traffic, or on other runways if there are more than one.

By regulation, right-hand approaches at fields with a left-hand pattern are *verboten*. It is senseless to be out of step with everyone else, but every so often we see someone try it, usually attempting to save a few minutes of rental time. You may think you are the only airplane out there, but remember what we said in the previous section: It's the one you *don't* see that gets you. A straight-in approach, on the other hand, is quasilegal, but it's dumb, because an airplane on a long final can be nearly invisible to other traffic.

It may occasionally be convenient to make a base-leg entry to an uncontrolled pattern to avoid a 180° turn on the downwind leg. Such an entry does assure at least one turn before turning final, giving a measure of visibility. Never, however, make a base-leg entry if there is traffic on downwind. In addition, my old daddy taught me to look both ways before crossing a street, so I always look out the right window before turning final, in case someone's out there dragging it in.

Runway manners are every bit as important at an uncontrolled field as at a controlled airport. You should plan on clearing the runway promptly after touchdown, so land and roll out in a manner that takes advantage of any exits that were noted before landing. In the absence of a taxiway, most pilots will turn around to taxi back on the runway, so leave time for the preceding aircraft to back-taxi and clear the runway. If you are talking to another airplane and want to make arrangements with that pilot about landing short or rolling out long, fine and dandy, but in the absence of such a scheme, follow another plane with your hand on the throttle, ready to go around.

If forced to make a waveoff, climb out beside the runway so you can see other aircraft and ascend back to the downwind leg as if taking off. Don't try a tight, low, 360° turn back onto final; the residents won't like it, the other pilots may not see you, and if the engine quits you'll be in a hazardous position. On the other hand, don't fly a superextended pattern just because there's another airplane ahead of you. Use only enough airspace to give breathing room. Generally speaking, a half mile is close enough when following another airplane, and that assumes you can match speeds. A proper traffic pattern size gives a final approach of ½ to 1 mile in length;

a 2-mile final is a bit wasteful of fuel unless it is required because of a string of airplanes landing ahead of you. It also leaves too much open space, tempting other aircraft to unknowingly turn inside you on a closer base leg.

USING THE RADIO

The proper advisory frequency is the charted CTAF, usually 122.7, 122.8, 122.9, or 123.0, used for position reports even if no one answers. Always start and end your call with the airport name: "Podunk traffic, this is Cessna one-four-two-five Victor, turning downwind for one-two, Podunk." In this manner, inhabitants of other traffic patterns won't have to listen to your entire call to learn your location and those who tuned in late get another chance to hear the name. Calls on base leg, final approach, taxiing back, and clearing the runway should be made only if you don't walk on someone else's transmission.

When parking, try not to block other aircraft, taxiways, or the fuel pump area. Unless there are acres of ramp space, always ask someone where to park before you leave for town, even if you think you have chosen a vacant spot.

Life at small country airports doesn't have to be traumatic, but unfamiliar pilots must remember that they are strangers in a strange land, and they must keep an eye out for the unexpected. Take a little more time than you might when flying in a controlled airport environment, making sure you are conforming to the flow of traffic and aren't getting in anyone's way. Even the busiest uncontrolled fields can be safely negotiated if you just don't insist on having everything your own way.

GRASS STRIPS

Creating an airport can be about as simple as measuring out a plot of vacant pastureland and informing the FAA that you would like the meadow listed as an approved landing facility. *Approved* for private-use purposes means little more than that the field was once usable in some fashion and that it posed no hazard to existing airports. Thus it comes as no surprise that of the 15,000 U.S. airports, only about one-third have paved runways.

For those of us who grew up, aeronautically speaking, with pavement under our tires, landing on grass can be a refreshing change—or perhaps a bit of a shock if the field is poorly maintained. Grass fields can vary widely in their suitability for aircraft use, and the same runway can change radically with advancing seasons. Before you flight plan into a strip known only as a magenta circle on the chart, take time to review some grass field facts.

WHAT ARE "GRASS" FIELDS?

First, that beckoning runway may not be grass at all; the open circle symbol means only that the strip is unpaved (or has pavement less than 1,500 feet in length). The surface could be carefully groomed bluegrass, native prairie sod, cinders, packed sand, or just plain dirt. Unless you have reliable information of a recent nature, treat an unfamiliar turf airfield as possibly unsafe, and have a paved alternate within range should it prove to be unusable when you arrive. Even such an unknown factor as a 3-inch rainfall the night before could prove to be your undoing, should you touch down on the soggy surface.

Naturally, the type of aircraft used will influence your decision about a turf field. Taildraggers generally have a larger measure of utility on such strips, because they tend to have lower touchdown speeds and less landing gear resistance on takeoff. Light twins, on the other hand, should be used only on the best-maintained grass runways. The ideal turf runway is well-drained to dry out quickly after a shower, evenly graded to minimize humps and hollows, and covered with a dense, cultivated sod. If the grass is kept down to 2 inches in height and rolled frequently to compact the soil, no finer experience exists than sinking into the carpet with scarcely a sound from the tires. It beats the "chirp" of expended rubber any time.

HOW TO USE THE GRASS RUNWAY

Allow 50 percent more distance for operation from a grass runway in average conditions; if the vegetation is over the tops of the tires, double or triple the takeoff distance shown in the handbook. Landing is generally less troublesome, because the drag of the grass works to your favor, but brakes can be rendered inoperative by tall grass wrapping around the wheels, and wet grass can be as slick as icy pavement. Stopping is also made more difficult by the aircraft's vigorous bouncing on rough sod, because the wheels are only in momentary skidding contact with the surface.

Successful operation from a sod field will depend on your familiarity with rough-field procedure. Most turf strips are anything but smooth, so land with the last ounce of airspeed used up, remaining aloft as long as possible before subjecting the landing gear to the punishment of the rough

Because grass strips look like the rest of the countryside, one must be certain to land in the designated area, and not off the runway itself. Here we're about to land on a 50-foot-wide grass runway between rows of field crops and a road—an easy decision.

surface. The nosegear should be held off until elevator control is lost and the stick kept full back, even after the nose comes down. Use brakes only as needed to keep excess weight off the nosegear.

Takeoffs should be made by transferring the aircraft weight to the wings as soon as possible, relieving the landing gear of stress before the speed builds up. If flaps are available to increase lift, use the recommended maximum setting. If the nosewheel is not lifted out of the grass early in the run, normal takeoff speed might be impossible to achieve within the limits of the runway. So, full aft stick should be used to bring the nosewheel off, then back pressure used as needed to maintain the nose-up attitude. As soon as the aircraft begins to feel light, a slight amount of additional back pressure on the stick will bring liftoff. Follow with an immediate level-off into a shallow climb, gaining speed before leaving ground effect.

Two seasons of the year call for extra caution around turf fields: early spring and early summer. During the last weeks of winter, daytime thawing and nighttime freezes keep moisture near the soil's surface, making the runway unusable except while frozen in the morning hours. Early summer brings the hay season, when many private farm strips are allowed to grow knee-high so the owners can harvest the hay and seed crop.

IS THE FIELD OKAY FOR USE?

So much for the art of handling the airplane on grass, but how do you judge the field's condition before use? Walk the entire strip if at all possible, checking for new gopher holes, muddy spots, and rocks. The limits of a grass runway should be marked in some manner to keep pilots away from ditches and other hazards. Permanent markings can be half-buried tires, reflectors, or wooden markers on posts. Temporary flags on short stakes can also be helpful. Nothing should be used that would be a hazard to the airplane if struck accidentally.

If the surface proves reasonably solid and free of obstructions, and you decide to give it a try, taxi at reduced speed to lessen the shock to the landing gear and airframe. If you inadvertently bog down in a muddy spot, use full aft stick and lots of power to keep moving, so the tires will remain afloat long enough to reach solid ground. Once stopped, the tires sink deeper and deeper until escape is impossible. Small nosewheels, when allowed to sink into the mud, give the propeller a downward thrust line that turns the airplane into a mole; I have seen nosegear fairings burrowed completely out of sight.

If you are aloft, the runway's condition is naturally more difficult to ascertain. Circle overhead at pattern altitude to look for bare touchdown and rollout tracks, a clue to the best area to use. Note wet areas, marked by darker-colored soil and grass. If a breeze is blowing, watch for ripples in the grass like waves on a lake; grass tall enough to ripple is too tall to land on.

When the grass is worn or nonexistent, look for ruts in the dirt, so you can avoid those bone-jarrers as you touch down. Washouts will show up from the pattern, as will terraces and waterways. After the high recon, make a low pass down the runway at 50 to 100 feet to drag the field, keeping well above stall speed but slower than cruise. Look for the smaller holes unseen from higher altitude, and watch for standing water shining beneath the grass cover. Then apply full power and climb back to normal pattern height for your approach, if all looks well.

MAKING CONTACT

Because of their unimproved nature, grass fields seem to have more than their share of obstructions on the approach lane, as well as vagaries in their terrain. The obstructions are best handled by a steep, full-flap approach (flaps are needed for the soft-field touchdown anyway). Sloping terrain is another matter, affecting both approach and touchdown.

If the runway slopes uphill, the approach will look high when it really isn't, and you may wind up needing a last-minute shot of throttle to clear the fence. More-than-normal back pressure is needed to flare into the

This town's grass strip is a barely visible track in the prairie at the right edge of the picture. Care must be taken to land only in the designated area.

nose-up attitude demanded by the runway's slope, and the effective sink rate may be great enough to require power during the flare, but at least the rollout will be short with gravity's assistance.

Landing downhill means going around if touchdown is not made in the first couple of hundred feet of runway. The approach will be partially to blame, because the slope of the ground makes you look about right when you're actually going to land long. Practice those spot landings for such an occurrence, so you can get the tires planted as soon as possible, and don't be ashamed to open the throttle once again if the ground just keeps falling away from you during your attempt to land.

The rolling, undulating strip can be a problem for the unwary, especially if you neglect to hold the stick full aft as you contact a rise in the ground. It is best to touch down in the low spot just before ascending a rise so that you can decelerate on the way up and stay solidly on the ground as you go over the hump at the top. Pilots taking off from rolling strips should plan to lift off on the downslope, perhaps with an over-the-hill start to ensure lifting off before passing the low spot and encountering rising ground.

Finding a sod strip from the air may take some doing in itself, because grass fields look like...well, grass fields. The charted location will provide a place to start, answering such questions as which side of the railroad tracks, how far from town, or between which two roads the strip lies. Once you are sure of the general area, start looking for the long, open space required for a runway and the inevitable wind sock. In the absence of boundary markers, ascertain the center of the runway with care; look for a worn track and for obvious indications of mowing.

TAKING CARE OF THE AIRPLANE

Preparing the airplane for sod strip operation should include service to the shock struts; a soft strut is only a minor annoyance on paved runways, but a severe threat to the airframe structure on rough fields. Keeping the tires slightly underinflated can improve shock-absorbing ability and increase footprint area. It might be wise to remove wheel fairings, especially during wet weather, so they can't fill up with mud and frozen slush. Check the carburetor air filter every time you're outside the airplane; it will require frequent cleaning to remove grass seeds and clippings.

Summing up, landing on a good grass airport is a sensuous experience worth all the extra precautions. Even Learjets have been successfully operated from grass, so if the field is long and smooth enough, have no qualms about using it. Some of those unpaved strips are right where you've been wanting to go, and with a little respect and judgment you can probably use most of them.

TOWER-
CONTROLLED
AIRPORTS

Not everyone is fortunate enough to be able to learn to fly at a tower-controlled airport. After all, there are only a few hundred towers among the 15,000 U.S. airports. Thus, a lot of pilots gain only a nodding acquaintance with tower procedures.

In many cases, the newly rated pilot remains deathly afraid of doing the wrong thing at a controlled field and bringing down the full wrath of the Federal Aviation Administration as a result of his or her ineptitude. Therefore, the new pilot avoids busy airports like the plague, which only makes matters worse, because lack of use quickly dulls the meager skills acquired when fulfilling the licensing requirements. Confidence in any piloting operation comes from knowing in advance what is likely to happen and how you are going to make it happen.

There is, of course, no need to fear the controlled airport. The tower personnel aren't out to write tickets; all they want to do is keep the airplanes moving safely. Filing a violation is done only as a last resort when a pilot displays an utter disregard for cooperation and willfully creates a hazard to others. Tower personnel don't like the paperwork of violations and as long as the pilot is honest with them, minor miscues can be overlooked. So much for the fear of towers.

In some respects, using a field with a control tower is actually simpler than flying into a noncontrolled airport. The tower controller assigns runway and landing sequence for you, and you will probably be allowed to approach directly from your present position, be it straight-in, right-hand base leg, or whatever, rather than be locked into the standard pattern of a nontower airport. All you need do is acknowledge receipt of tower instructions and carry them out, and the more familiar you are with procedures, the easier this becomes.

Tower controllers like these aren't out to write tickets for violations. Their job is to keep traffic moving safely on and around the runways.

ARRIVAL PROCEDURES

Let's say you are headed for Smogville International, a little jittery about this first trip into the city for several months. Studying the chart while en route, you note the pattern of the runway layout, and look for the letters ATIS with a frequency, found under the airport name. If the airport has such an automatic terminal information service, listening on that frequency before announcing your arrival will help immensely. ATIS gives a recorded playback of weather and wind data, runway in use, and any NOTAMs (notices to airmen) critical to operation on the field. A code word lets the tower operator know you have listened to ATIS if you say "have information Kilo" or whatever the current password is.

If there is no ATIS, listen for a few moments to the conversations on the primary tower frequency (the bold one shown on the chart after the letters "CT") and you will be able to glean essentially the same information as ATIS. Then you can tell the tower "I have the numbers," to let them know to save their breath. Without some tip-off, the tower folks are honor-bound to give each new arrival full weather and traffic info; using ATIS or listening first makes a friend right off the bat.

Having listened to get the picture, look over your situation in light of the runway in use to anticipate what the tower will want you to do. If runway 17 is in use, and you are approaching from directly east of the field, it would be logical to expect a clearance to enter left traffic for 17, either a downwind entry or a base-leg entry. If coming in from north of the field, you might anticipate a request to report on a long final, straight in for 17. Mentally prepared, you are much more likely to understand a controller's rapid-fire delivery, as well as be on top of any misunderstanding.

In some case, your first contact might be with approach control, rather than the local tower controller, particularly at airports with Class B or C airspace designations instead of Class D. These extra layers of airspace involve using only one more frequency and checking in somewhat farther out from the airport; otherwise most procedures are the same. If approach control's assistance is called for, ATIS may give the frequency, or the control tower may suggest that you go over to approach on frequency such-and-such when you call in too far out. In a pinch, flight service will gladly provide it for you; approach control frequencies are also shown on the border of the chart, and are published in the *Airport/Facility Directory*. However, unless the airport lies within Class B or Class C, contact with approach control is optional.

Remember, a basic control tower has jurisdiction only over its Class D airspace, roughly defined as a 4-nautical-mile radius of the airport up to 2,500 feet above the surface. Cooperation on the pilot's part includes not running right up to the boundary before hollering "I want in." An advance call gives time for several blocked-out transmissions and pauses for a lull in traffic before actually violating the Class D area. Make your first report at some prominent point well outside the 4-mile limit so the controller will instantly know where you are. However, something like "seven miles east" will suffice if you don't see anything clearly marked on the chart. Be sure to listen briefly with the volume up before pushing the mike button, so as not to interrupt another transmission.

When making the initial call-up, keep it simple but complete. Tell 'em who you are, where you are, and what you want to do: "Smogville tower, this is Cessna one-two-three-six-Quebec, over the stadium, landing, with Kilo, over." You've said it all, the tower controller has only to decide what to do with you. The reply probably will assign a closer-in reporting point and an entry path to follow: "Cessna three-six-Quebec, report on left base to runway one-seven." Because you are not yet in sight, your sequence behind other aircraft cannot be assigned until you report on left base.

Acknowledge all instructions as received and understood by giving your call sign, abbreviated to the last three figures if the tower got it right on the initial call-up. Do not acknowledge unless you did understand; if you aren't sure what they said or what they meant, *now* is the time to say so, not when you're in the middle of a muddle of airplanes. Controllers will talk at a pace acceptable to the pilots they are used to working with and as a stranger you may not copy everything the first time. Tell 'em "Say again, please" until you do understand, then give them the "three-six-Quebec" acknowledgment.

After the first call-up you have only to fly closer to the airport, following the approach path requested by the tower. Be watching for other traffic as you hear them contact the tower, so you have the plane located before the tower gives them to you as traffic.

Reporting your arrival at the assigned point is a simple pronouncement: "Cessna three-six-Quebec, left base for one-seven." The controller now has you in sight and identified, and issues either a "cleared to land" or a sequence, such as "number two, behind the red Bonanza on one-mile final." Acknowledge all tower instructions, and if given a sequence behind another aircraft, let the tower know whether or not you see it. Say either "I've got the traffic" or "no contact." If you acknowledge having your traffic in sight, separation becomes largely your responsibility. If you report no contact, the controller will point out the other aircraft in greater detail, or will steer you well clear.

When following another aircraft, do not crowd in closer than 3,000 feet, or you may be asked to go around. On the other hand, don't waste everybody's time by using a 2-mile interval. Use your head by keeping your speed up when following a fast airplane until just before landing. If you know your traffic is going to be slow, plan on letting it all hang out (lowering flaps) to maintain your position, or S-turn slightly. If you are number one and you know you have traffic behind you, be a good neighbor and keep the pattern in close and reasonably fast.

You will normally be "cleared to land" only after the airplane ahead has turned off the runway (unless no conflict is anticipated), but as long as you have been assigned a sequence, continue your approach, expecting the clearance before you touch down. Should the tower remain silent, an inquiry might be in order: "Is three-six-Quebec cleared to land?" But if the controller is talking 3 miles a minute to another aircraft, go ahead and land as long as there is no conflict in view. Technically, you are in violation, but as long as a normal traffic flow is maintained, everybody's happy.

If you do feel a go-around is necessary, get the airplane organized first, then advise the tower as quickly as possible what you are doing: "Three-six-Quebec is going around, we'll try another landing." The tower may tell you "make left traffic, report downwind," or if the pattern is clearly defined, may just say to follow the aircraft ahead. Reporting on downwind is tantamount to asking for a landing clearance, so if you want to do anything different than the previous circuits, this is the time to advise the tower: "Three-six-Quebec downwind, this will be a full stop" advises the controller that you are finished touching and going, for instance. "What are your intentions?" is a request for information.

GETTING AROUND ON THE GROUND

After touchdown, head for a convenient exit as soon as you have good ground-handling control; don't come to a full crawl first and cause a go-around. The runway is hot property, so don't keep it any longer than is necessary. Stay on the tower frequency until turning off, then check in on the ground control channel unless advised not to by the tower. It is not necessary to stop while calling up ground control; as long as no conflicting traffic is coming at you, it's acceptable to let the airplane roll during the few seconds before contact. Hold your position if another airplane is approaching, of course. If you don't know ground control's frequency, fear not; the tower will probably tell you, "Contact ground control on point six when clear," or you can ask them. There are basically four frequencies in general use, 121.6, 121.7, 121.8, and 121.9, of which .9 is the most prevalent, but some newer towers or ex-military fields have odd ground control channels.

If you know where you are going, tell ground control what you're doing when you check in: "Three-six-Quebec, just clear of one-seven, going to Gulf Air." If you don't know where you're going, say so, asking for transient services or the terminal. The controller can then take you by the hand, leading you through the maze of taxiways to the ramp. Tower personnel cannot recommend parking or service facilities for you, so watch for flag persons or serve yourself as you see fit.

GETTING OUT AGAIN

After finishing your business, departure is equally simple. Listen first to ATIS for any changes since you arrived, then advise ground control of your location and desires: "Smogville ground, Cessna one-two-three-six-Quebec, at Gulf Air with Lima, taxi for takeoff, over." Switch on an anti-collision beacon or pull forward a few feet to help the controller pick you out of a line of airplanes. "Three-six-Quebec, taxi to runway one-seven via north on Alpha" may be the reply. Taxiways are commonly named by letters of the phonetic alphabet, so acknowledge, then look for a sign with the letter A on it. Yellow center strips mean taxiways, as do blue edge lights; don't trundle out onto a wide taxiway with white markings and clear lights, because that's a runway!

You are cleared to cross intersecting runways as long as ground control said "Cleared to runway one-seven." If the clearance was simply, "Three-six-Quebec, runway one-seven, hold short of two-four," be alert for an intersecting runway somewhere en route to one-seven, and stop short of it if you haven't been cleared to cross by that time. All hold-short instructions must be read back as given, to confirm that you understand exactly what to do. If in doubt, ask before crossing any yellow hold lines.

In some cases your initial call-up will be to a clearance delivery frequency rather than ground control, as advised by the ATIS. This extra step is part of the procedure to use at a Class B or C field wherein the approach controller, now called departure control, will assist in your departure from the area. The clearance delivery person will assign you an initial altitude to fly, a transponder code, and a frequency on which to contact departure control after takeoff. As with the use of approach control during the arrival, the procedure is optional at a TRSA (terminal radar service area) airport and may be declined by telling ground control, "Negative radar service," when you request taxi instructions. However, if you are only an occasional visitor to the airport, it is wise to use radar for both arrival and departure.

When arriving at the runway, a double yellow line perpendicular to the taxiway marks the "hold-short" line, which you are not to cross without tower permission. Do your runup off to one side of the taxiway so other pilots may pass you to depart if they're ready first. You may, of course, use an intersection short of the runway's end as the start of your takeoff roll, so long as you advise ground control of your intentions first, but remember, the responsibility is *yours* if it turns out you needed more.

Following the runup, switch to tower frequency and advise "Cessna three-six-Quebec, ready for takeoff on one-seven," or "midfield one-seven," if making an intersection departure. It is common courtesy to advise the tower of your direction of flight, such as "straight out to Blytheville," and the tower may solicit this information.

You may receive an immediate "Cleared for takeoff," you may be advised to "Hold short," or you can be told to "Pull into position and hold." The latter means to get out onto the runway and line up, but to wait for a "Cleared for takeoff" while traffic gets off the runway. If the tower says "Make an immediate takeoff or hold short," it means haste is needed because of landing traffic, so if you elect to go, get moving first and acknowledge on the roll. After you break ground the tower will advise you when to switch to departure control, if you're proceeding with radar service. In any case, monitor the last assigned frequency until clear of the facility's airspace.

Remember, control towers are provided as a service to the pilot, so don't be afraid to venture near the busy patterns. Use these fields, and soon you will find that picking up the mike and asking for a clearance is as simple as talking to Unicom. Practice makes it easy.

CLASS CHARLIE AIRSPACE

Once upon a time, in the days before radar and widely varying traffic types, tower-controlled airports were content to manage the traffic in the immediate proximity of the airport, within the controller's range of vision. With the advent of terminal radar coverage, however, the ability to sequence traffic at greater distances became a reality. The result was something called Stage III radar service that became an ARSA, or airport radar service area, later called Class C airspace. This is a piece of airspace roughly 10 nautical miles in diameter and 4,000 feet high surrounding a moderately busy airport that is typically used by all classes of aircraft and pilots, creating an interesting traffic mix. A little familiarity with the concept will make flights in and out of Class C airspace simpler.

Class Charlies are, quite simply, enlarged Class Delta areas, typically the old 5-mile radius (although nautical instead of statute) plus another 5-mile ring with a floor at 1,500 feet that permits nonparticipating airplanes to skim along in Class E airspace under the overhanging Class C. The top of the entire mushroom is usually set at 4,000 feet AGL. Although altitude-reporting transponders will be required to fly in or over the Class C, no extra certification or flight planning is needed. Unlike the less-stringent TRSA, or terminal radar service area, contact with ATC is not voluntary: "You VILL participate, and you VILL like it!" In theory, one should never encounter an unknown, nonparticipating aircraft inside Class Charlie airspace.

Assume you are inbound to an airport, and, glancing at the sectional chart, you find that two magenta rings surround the airport symbol, and sure enough it's listed in the table on the margin of the chart as a Class C airport. In all cases, an ATIS (automatic terminal information service) broadcast gives advance airport information, free for the listening. The ATIS frequency is found on the map in the airport's data block.

After listening to the ATIS, we glean the code word from the end of the sequence and dial up the approach control frequency, which is often included on the ATIS recording. It's also found in magenta rectangles outside the Class C rings. The sectional chart also includes approach frequencies in a tabulated listing of airports with radar service. Failing this, the local FSS specialists can look it up for you if you'll contact them.

"LANDING REGIONAL"

Referring to the chart, we note a magenta pennant with an underlined black place name nearby. These designated reporting points, familiar to the approach controllers, are fast tickets to enter the Charlie airspace, so it is helpful to defer a contact with approach until reaching one of these spots. Entry to the Class C is permitted only after establishing contact with the controller, and this merely means the reply must include your N-number.

Therefore, your call-up, "Regional approach, Cessna seven-zero-four-Sierra-Juliet, over Pittsville with X-ray, landing at Regional," will probably evoke the reply, "Seven-zero-four-Sierra-Juliet, squawk four-six-two-zero." At this point, you have established contact, satisfying the letter of the regulation, even though radar identification is still forthcoming. If the controller says, "Aircraft calling approach, stand by," you have *not* made contact, and must stay well clear of the Class C boundary.

Airports such as this one, surrounded by Class C airspace, require contact with approach control when arriving and departing. Consider them to be merely larger chunks of Class D airspace and use them with proper preparation.

If the controller sees your target on his or her display, you'll be told, "radar contact," and your location confirmed. You will probably be asked to verify your altitude to make sure it matches the one shown on the radar display. If you're handed off to another controller as you move along, your check-in should include your altitude for verification, as in "four-Sierra-Juliet, descending through three-thousand three-hundred." Approach control will assign you to an altitude and heading in many cases, unless you happen to be the only aircraft aloft in your section of the Class Charlie (it happens quite frequently). However, this "hard" course assignment does not mean you are relieved of observing VFR cloud clearance regulations; if you will be unable to keep the required separation, take evasive action and inform the controller of your needs.

Approach's job is to sequence you for the control tower with the proper spacing from other aircraft and to keep you separated from conflicting traffic. You'll be told which runway to expect and how you're going to get there, subject to change by the tower. Acknowledge all instructions with your abbreviated call sign, reading back the important numbers like heading, altitude, and runway, and if you don't understand, ask for clarification.

Traffic callouts are given in clock positions, with 12 o'clock being your aircraft's path ahead. If approach says "traffic 10 o'clock, two miles, a King Air climbing through three thousand," your job is to look at your left front quarter for the climbing Beechcraft and to tell the controller what you see, either "four-Sierra-Juliet has the traffic" or "no contact." If you report the traffic in sight, separation becomes your job; if you lose sight of it tell the controller "four-Sierra-Juliet has lost the traffic."

At the appropriate point, usually a few miles from the field or when you have reported it in sight, you will be told to switch over to the tower frequency. Make the check-in brief, because the tower knows you're coming. Procedures from this point on are similar to any tower-controlled airport, with perhaps a bit more emphasis on "keep your speed up on final, jet traffic is 'marker inbound.'"

"READY TO COPY"

As we said in the previous chapter, **when you're ready to leave, the departure requires an initial contact with a person addressed as "clearance delivery,"** with a discrete frequency. Before moving from the parking spot, contact clearance delivery for a VFR clearance out of the Class C airspace. They will want to know your destination or on-course heading and requested cruising altitude; you will be assigned a transponder code, departure control frequency (which is usually the same as approach control), altitude limit, and sometimes a departure heading, although the heading is most often assigned by the tower controller when you are cleared for takeoff.

Write the numbers down on a notepad for later reference and read the elements of your clearance back to verify correctness. Don't forget to dial the new code into the transponder, which you should have left on "standby" because it contained an older code that was no longer assigned to you.

Once the clearance is in hand, contact ground control in the usual fashion. Taxi and takeoff procedures are normal for a controlled field, but you will be following a hard altitude and heading assignment once you are airborne. The control tower will hand you off to departure by the time you reach the airport boundary and once you're in radar contact you'll just follow the controller's instructions. Keep track of your position, because you may be vectored far away from your direct route until clear of arriving traffic, at which point you'll be told "cleared on course." The departure controller will provide service well beyond the boundary shown on the chart, into an outer area to the limit of the radar coverage, unless you ask for a "frequency change," meaning you want to drop out of his or her service and squawk 1200 on the transponder.

If you are operating to and from another airport that happens to lie inside the Class C circles, you must either stay below the floor of the outer ring or contact approach/departure control for flight through the Class C airspace. Don't crowd the limits, particularly if encoder equipped; the altitude readout jumps in 100-foot steps, and the readout might round off your altitude to the next higher digit. You should keep a minimum of 200 feet of room below the base of Class C if you're skimming along underneath, and that will leave you uncomfortably close to the 1,000-foot minimum flight altitude over congested areas. If departing at an uncontrolled fields inside the inner ring, you are required to check in with departure control as soon as you are airborne, heading away from the primary airport.

Class C procedures need not be intimidating. Just know where you're supposed to go and be ready to cooperate. We can't change the requirements, but we can learn to work with the system we have.

CLASS BRAVO AIRSPACE

"Hoo-boy, I'm not going anywhere near *that* place; they've got Class B airspace over there!" This fearful attitude keeps many perfectly capable pilots from using an airport surrounded by what used to be called a terminal control area, all because they feel inadequate to cope with the procedures involved. Actually, entering a Class B airport is not very different from going into any busy field; some optional procedures simply become requirements in Class B.

The original Class Bravo concept, called a terminal control area (TCA), was to designate the nation's most busy fields as Group 1 TCAs and less busy airports as Group 2, where solo student pilot flying would be permitted. Thus, not all Class B airports are superbusy. The Group 1 and Group 2 designations were dropped in 1989, but some student training is still permitted at the former Group 2 locations.

CLASS B AIRSPACE DEFINED

Think of Class B airspace as a giant block of Class D airspace, with everyone, IFR or VFR, given radar separation and sequencing to the same standards. There's no difference in pilot procedure between an airport with a terminal radar service area or Class C airspace or Class B. In fact, with TRSA service you must keep a sharp watch for nonparticipating traffic when receiving radar advisories, a hazard lacking in Class B where everyone is under radar control. Nor do you have to worry about staying 500 feet below or 2,000 feet away from clouds to fly VFR in Class B airspace; the rules only require you to stay clear of clouds, because no unknown IFR traffic is going to appear suddenly out of the murk.

Prepare for Class Bravo by requesting and utilizing radar service at some less-congested fields. In this way, you learn to interpret the radio communications dealing with traffic advisories, vectors, and clearance assignments. Remember, at these airports more radio frequencies will be used. Look them up and jot them down in the order of use before taxi, and leave room to note clearance details and additional frequencies assigned in flight. The ATIS, or automatic terminal information service, will be the first channel used, enabling you to pick up the recorded weather, traffic, and NOTAM information. You can listen to the ATIS as many times as you like, but don't do what many pilots do and listen only long enough to pick up the code word without checking the rest of the broadcast for traffic information and special notices. You might miss something important, such as a taxiway closure or a frequency that has been changed.

From ATIS you can anticipate your arrival path, and also which of the instrument approaches is being used—important information even to VFR pilots, because the type of approach (VOR, GPS, ILS, visual) can tip you off as to the arrival procedure being used by jets and heavy aircraft. Check your altimeter setting to make sure your cockpit indication will be the same as your encoded readout on the radar display, and note the crosswind component you will encounter during landing. Then verify that you will be using the appropriate approach control frequency for the initial contact.

Most busy airports have sectorized approach control, with separate frequencies for various arrival quadrants, and possibly different channels for VFR versus IFR arrivals. Never fear; if you should call on the wrong fre-

Class B airports, typically airline hubs like this field, require extra careful procedures from pilots. They are, however, no more difficult to enter and exit than any busy tower-controlled field, and should not be written off as "untouchable."

quency, you'll be told to switch to one-two-one-point-five-five or whatever. Always listen for a few seconds before pushing the button to make sure you aren't breaking into a conversation, and have your spiel ready for delivery so you won't have to compose it while holding down the mike button.

WHAT TO SAY

Every ATC controller is interested in three things; *who* you are, *where* you are, and *what* you want to do. Try to get these three items across in as little time as possible and you'll be cutting down on frequency congestion. Pick out a prominent reporting point for your contact, such as "over Perryville reservoir," or "over the VOR" (name the station if there is more than one). Buy a Terminal Area Chart, the one with a scale of 1:250,000, and look for pennants with underlined place names; these indicate VFR reporting points, not shown on the sectional chart. A charted VOR airway intersection is known to ATC, of course, and also makes a good reporting point. Begin trying to contact the approach controller well outside the 30-mile "veil" ring surrounding the Class B airspace, giving yourself time to be acquired as "radar contact" *and* to receive a clearance before you hit the Class Bravo perimeter and become illegal. This also gives you an opportunity to make sure the transponder is working, because it is a requirement to fly inside the 30-mile Mode C veil, even outside Class B airspace.

Your first call will be something like this: "Downtown approach, this is Cessna one-two-three-six-Quebec over Backwater with Oscar, landing Downtown, over." In one short sentence you have told the approach controller most of what he or she needs to know; the response can be "Cessna one-two-three-six-Quebec, squawk one-six-two-seven," the latter being the discrete transponder code which is next in line in the stack for assignment. With the assignment of a code, you are tagged as a specific airplane, complete with N-number and altitude. Acknowledge the transmission with your abbreviated call sign and read back the code as you dial it into the transponder. The next transmission you hope to hear will be "Cessna three-six-Quebec, radar contact 2 miles northwest of Backwater airport, cleared to enter the Class B airspace." This signifies the controller's ability to provide the necessary separation and sequencing for Class B entry. Without hearing these words you are *not* to enter the Class B airspace!

Even though you are equipped with an encoding altimeter, the approach controller will ask you to "Verify three-thousand six-hundred," just to make sure you're in agreement with the computer's readout. Approach also has a groundspeed tagged to your target to help them sequence various types of aircraft.

If asked to "Squawk ident," your reply should be made through the transponder by simply pushing the ident button; no radio reply is needed because you are acknowledging receipt of the instruction by painting as an "ident" return on the radar display. If a second request to ident is made, you can tell the controller you've already idented, in which case you might have a malfunctioning transponder that isn't showing up on the display. It is possible to negotiate an entry without an altitude encoder if you have suffered a failure en route, but a basic Mode A transponder is usually necessary unless prior arrangements have been made.

BEING LED BY THE HAND

After radar contact is established, approach control will assign a heading, or radar vector as it is called, to sequence you with other traffic and an altitude to maintain that will assure 500 feet of vertical separation. The transmission might be "Cessna three-six-Quebec, turn right, heading one-four-zero, maintain at or below three thousand." This means you can use any altitude below 3,000 feet MSL, but you must not go above 3,000. It is well to read back the instructions unless the controller is already turning his or her attentions elsewhere, but keep it brief, such as "Three-six-Quebec, right to one-four-zero, descending to three thousand." In this way, any misunderstandings will be corrected before you wander too far afield. I might add that it is wise to check your directional gyro for precession before receiving radar vectors; the controller will have enough trouble allowing for a crosswind on your track without tossing in 10° or more of precession error between the directional gyro (DG) and the magnetic compass.

Jot down the frequency being used when time permits. This gives you a record of the last controller's channel in case you have need to recontact later. Even with flip-flop digital tuning one can get confused. Although the control tower will probably be the next voice you'll hear, handoffs are sometimes made if you cross the division line between approach sectors. Whenever you're told to contact a new controller, read back the change by saying "Three-six-Quebec, one-two-four-point-seven" with a brief pause afterward to see if the controller makes any comment; you may have heard wrong. Dial it up promptly, rather than jotting it down first, and listen for traffic a second or two before reporting in to the new controller. Your contact will be simply "Downtown approach, Cessna three-six-Quebec, three thousand," and the reply will be a simple "Roger" unless the new controller has further instructions. If no response is heard after a couple of tries, go back to the previous frequency you had noted earlier and tell that controller you're unable to contact the other controller. As soon as contact is made, write down the new frequency.

Traffic advisories will be issued as needed whenever a conflict is apparent. If the controller says, "Three-six-Quebec, you have traffic at 10 o'clock, westbound, three-thousand five-hundred" scan the horizon between the left wing and nose; you must allow for a crosswind's effect on your heading. Look high and low, then report your findings, either "Three-six-Quebec, no contact" or "Three-six-Quebec, I have the traffic." If you say you have the traffic, keep it in sight until it is no longer a factor, and if contact is lost before the aircraft is clear of your path, ask ATC for its position. Even in Class B airspace, where IFR separation standards are maintained, the final responsibility for separation is always the pilot's. If you say, "No contact," the controller will monitor the situation and issue updates until you spot the bogey or until the controller can report that it is no longer a factor. Do not substitute the cabin speaker for your eyeballs. Maintain your own watch even while operating in a radar environment, and enlist the aid of your passengers as well.

SWITCHING OVER TO THE TOWER

With successive heading changes, the approach controller will align you with the arrival gate as desired by the tower controller, and will tell you when to switch to the tower frequency, usually only a few miles from the runway. You will be asked to verify that you have the field in sight. "Three-six-Quebec is four miles northwest of the airport, contact tower now on one-one-eight-point-four" will be a typical handoff. You will sign off with "Three-six-Quebec, going to tower" and switch to the tower frequency you should already have tuned into your number two frequency position. You have only to report your location, because you are being expected by the tower controller: "Downtown tower, Cessna three-six-Quebec, four northwest." You will be issued a traffic sequence, such as "Number three, follow the 737 on two-mile final," unless you happen to be the only soul in the sky at that moment.

More often than not, you will be sandwiched between two jets on final approach, and it will be helpful to all concerned for you to maintain as much speed as possible until just before flaring out. A 120-knot final is much more neighborly than a 70-knot one at a big airport. Practice these high-speed arrivals at home and you will soon learn just how close to the threshold you can come before bleeding off power and applying drag from the flaps and gear to touch down on a chosen spot. Plan your touchdown and rollout in order to arrive at a turnoff point without tying up the runway. If no taxiway exit is handy for some distance down the runway, aim a bit long rather than putting it on the numbers, letting the aircraft roll without delay to the turnoff and clearing the runway promptly for the sake of that jet on final. A delayed touchdown also avoids wake turbulence from the heavy aircraft that just landed, by keeping you above its flight path. Remember, the ultimate authority for safe operation of the aircraft belongs with the pilot, not the controller, so don't let a request for an expedited turnoff cause you to fold up the airplane in an attempt to comply. Just do the best you can.

Upon clearing the runway—not before—switch over to ground control frequency, which you should have noted on your flight log or placed in the standby slot, in case the tower is too busy to give you the usual "Three-six-Quebec, turn right next taxiway, point-nine clearing the runway." Again, tell the controller your ident, position, and intentions: "Downtown ground, Cessna three-six-Quebec, just cleared one-two left, taxiing to general aviation parking." For your first visit to this virgin sod, it would be well to have someone along who is familiar with the lay of the land; if this is not possible, get a good briefing on the ground layout in advance and have an airport diagram in your lap. You can solicit help from ground control by saying you're unfamiliar with the airport and the controller should be able to give you progressive taxi instructions. Remember, taxiways are named for phonetic letters, such as Alpha, Delta, Romeo, and so on, so you will be told "North on Delta, east on Romeo at Delta 1" as taxi instructions. Check your airport diagram and make sure you're turning up the correct taxiway.

GETTING OUT AGAIN

Departure from the Class B airport is even easier than arrival; once airborne the pace gets easier. Listen to the ATIS first, then contact clearance delivery on the frequency specified by the ATIS recording, not ground control. Clearance will want to know your direction of flight and desired altitude, so you can say "Downtown clearance delivery, Cessna one-two-three-six-Quebec is requesting a VFR clearance northbound to Armonk at three thousand." Clearance delivery will then assign the transponder code, an initial altitude limit, and frequency on which to contact departure control. Once you've read back the instructions to clearance delivery to verify you heard them right you can contact ground control, giving your position and intentions: "Downtown ground control, Cessna one-two-three-six-Quebec is at general aviation parking, ready to taxi for takeoff, we have Tango." Again, ask for progressive taxi instructions if you're not sure of yourself, to avoid embarrassment.

After runup is complete at the end of the runway, verify that the transponder is on and tell the tower you're ready to go. If you are attempting to speed things up by departing from an intersection, you may be thwarted by mandatory 3-minute holds that are due to wake turbulence from departing jets. It sometimes pays to taxi to the far end of the runway, where you may have the option of waiving the hold if you think you can avoid the turbulence. The tower will release you on a heading suitable to departure control's needs, such as "Fly heading zero-six-zero." ATC's worst nightmare is a no-radio airplane blundering into a line of traffic, so you will be kept out of the way until you are solidly in the system.

Shortly after liftoff you will be told to "contact departure control now on one-one-niner-point-three." You will check in with departure control as "Downtown departure, Cessna three-six-Quebec, just off one-two left, out of one-thousand two-hundred." Departure will reply "Three-six-Quebec is radar contact, maintain three thousand, cleared on course." Radar vectors and traffic advisories will be given in a manner similar to arrival, until you reach the outer limits of the Class B airspace or the fringes of radar coverage. At this point you will be told, "Radar service terminated, squawk VFR, frequency change approved." You reply, "Three-six-Quebec, squawking twelve hundred," perhaps with a word of thanks for the assistance. Be sure to clear your transponder by flipping the digits back to 1200 so you will not be squawking a discrete code that is assigned to another airplane.

As you can see, Class B entry and exit procedures are no more complicated than using radar services at a TRSA or Class C airport, although the pace may be somewhat quicker. You have only to follow instructions and keep on top of the situation, complying in a professional manner with the controller's instructions. Talk clearly and concisely on the radio, as the pro pilots do, but don't be so rapid-fire with your cryptic replies that you are constantly being asked to "Say again." Laconic replies also are frowned upon, so try to strike a happy medium.

Don't fear the big airport; just behave yourself and with practice you'll find that flying into Class Bravo airspace is almost like operating at home base.

WEATHER

PREFLIGHT
BRIEFING

The pilot was marooned by low stratus and fog for a couple of days, made frequent trips between the airport and the motel, and called flight service regularly for more weather information. I bumped into him at the airport coffee pot, and chatted sympathetically about his chances for getting out that day. I opined that the stationary front was bound to be drifting eastward soon, but a low-pressure area was impeding its progress and was pumping in a southeasterly flow of moist air. His response was a bit unexpected: "Where is the low?" he asked.

Now, we've all been in this man's shoes. Trapped away from home by a stalled weather system, we desperately want to know when it's safe to plan a departure at the earliest possible moment. Yet after 2 days of calling the FSS regularly, this pilot didn't know the overall weather pattern. Somehow, he never learned to brief himself, even with the assistance of flight service.

BEFORE YOU CALL

To safely fly cross-country with a minimum of inconvenience, you must know as much about the weather you're faced with as the person answering the FSS phone. True, the FSS individual is a trained weather briefer, and will do his or her best to give you what he or she and the briefer's manual believe is adequate information for the flight, but only *you* can decide whether it's good enough for you to go. You are a pilot; you will see more weather firsthand in a month of flying than most weather briefers will see in a decade. You should learn from each flight, and be better prepared for the next one by observing the conditions encountered.

Before calling the FSS, learn as much as possible about the overall weather situation, so you can be aware of problems and ask specific questions. You should know the location of highs and lows, as well as frontal zones, and the type of air mass involved (dry, moist, polar, maritime, etc.). Thus prepared, you can ask intelligent questions of the briefer and weigh the merits of the forecasts.

Finding out the big picture isn't all that difficult; one of the best TV presentations is still the early-morning *Today* show on NBC because it's made up only an hour or two before show time. Locally originated TV weather shows, although varying widely in content and veracity, can be better than nothing. If you are in a location with cable TV offering the nationwide Weather Channel, originating in Atlanta, it provides aviation weather periodically. If no prior information is available, make sure you get an area synopsis from the briefer, so you can relate it to the other data.

With access to a personal computer, pilots can have a wealth of weather information at their fingertips. There are any number of weather sites on the Web, mostly free for the taking. It takes a bit of skill to find your way through the plethora of options, and everyone has favorite sites and techniques. For the less computer-adept, it is not necessary to wade through the morass of Internet options; the FAA-sponsored DUAT (direct user access terminal) service provides pilots with dial-up access to the same aviation weather data that's available to the flight service station briefer, and it can be reached anywhere you can connect your computer to the phone system.

Unlike the jungle you must hack your way out of on the Internet, the DUAT providers (there are presently two competing for your business) host the National Weather Service's data and give easy-to-read menus to follow. Whenever you dial in for a weather briefing (it's more of a weather search, because nobody is there to brief you), DUAT is paid a small fee by the FAA. You must be a pilot with a current medical to be issued a secret account number, confirmed by a password, and an aircraft N-number will be required. Once in, you can receive a lengthy standard briefing, a route briefing, or just ask for specific weather products at specific locations. If you work online, move quickly or you'll be disconnected after a generous allotment of time.

Full-service FBOs (fixed-base operators) will have a pilot briefing area with a dedicated computer hooked up to a satellite feed, through which you can access commercial weather services on their nickel (it costs them a pretty penny to have this available). Don't hog the system if there are other pilots waiting. Like DUAT, you must know what you're looking for: the big picture, radar views of precipitation, current and forecast conditions along your route. The problem with any self-briefing is overlooking details, either because you're not a professional weather person or because all the information isn't available on your system.

Before you go, therefore, you should make a contact with a flight service briefing to update your knowledge of the overall weather situation. He or she may have just spoken with a pilot in the area you're heading toward,

The Weather Channel is an excellent way to gather information for the Big Picture, and then add your observations to it when you're forced to create your own weather briefing.

fresh pilot reports may have popped up and, most important, the FSS will be familiar with the day's security situation. Temporary flight restrictions can be designated with little warning and the NOTAMs (notices to airmen) are often difficult to find on your own. You don't need an F-16 fighter escort on your wingtip because you didn't know the president was in town.

NOW YOU'RE READY

For the most current reports of weather, call the FSS about 5 minutes after the hour. Weather observations are taken just before the hour, and for several minutes the computers are in the process of loading fresh reports, so calling just before the hour will get you reports nearly an hour old. Delaying a few minutes will get you the latest poop, perhaps altering your decision. Remember, however, that weather reports are only spot observations; they must be used in conjunction with the big picture to update the changing weather pattern. They are also largely generated by automated observation stations, and even if augmented by human observers they can miss approaching storm clouds. By gleaning information such as wind shifts at various stations, or changing cloud heights, you will be able to stay abreast of the situation.

Forecasts are important, naturally. They are prepared for several hours at a time, overlapping with the preceding forecasts for continuity. Area forecasts predict the changing big picture, and terminal forecasts take into account micrometeorology at specific airports, so you'll know if you can see to land after you get there. Use both, but compare forecasts diligently with reported weather to see how things are going. Believe no forecast until it has been proven true.

When you dial up the FSS, state your aircraft number and intentions for their benefit, then request specific information, rather than casting yourself on the briefer's mercy. Where is the front now? Have you had any pilot reports along my route? What was the last radar report on precipitation areas? Ask for reports and forecasts for the stations you consider significant, if you don't get them as part of the standard briefing. Reporting points "upstream" in the path of weather movement can bear a clue to the conditions moving into your route during the next few hours. Learn where the good weather lies, so you can head in the proper direction if your destination is unreachable. Don't settle for a canned presentation.

Above all, write down the information. You may have need of comparing the next hour's weather with this hour's to establish a trend, and unless you have a photographic memory, you won't remember the cloud heights 60 minutes from now. Devise your own shorthand system if need be, but note the weather reported at each station somehow for later reference.

ANALYZE WHAT YOU'VE HEARD

Exactly what are you looking for in this mass of data? You want to know what the current weather is along your route, and what it is forecast to be. In most cases the forecast is already in effect, so compare actual weather with the forecast. If the forecast was for improvement, has such a trend really begun? If the situation is supposed to deteriorate, is it proceeding as predicted? If improvement is slower than forecast, perhaps you shouldn't place too much faith in the forecast. On the other hand, improvement may be occurring well ahead of expectations, a welcome sign. Be especially watchful of winds that are stronger than forecast; these can indicate a low-pressure system deepening or moving in sooner than expected. In such a case, all bets are off, and amended forecasts probably will be out shortly.

Should you be faced with marginal conditions, try to establish a trend over several sequence reports, and stay on the ground until an upward trend is shown over at least 3 hours. Check several stations, to make sure you are not dealing with a localized problem. In any case, suitable alternatives should be within easy reach before you tackle weather that pushes your limitations.

As the pilot in command, you must decide for yourself if the weather is adequate for the flight. Safe VFR is not the regulatory 1,000 feet of ceiling and 3 miles of visibility; it is whatever ceiling and visibility will allow you a comfortable safety margin to hunt an alternative field if deteriorating conditions are encountered. Likewise, 200 feet and one-half mile are not adequate landing minima for a rusty instrument pilot. Marginal weather is defined by the limitations of your aircraft and your personal abilities, *not* by the FARs.

For the foregoing reasons, you cannot place the go/no-go decision in the hands of the FSS, asking in effect, "Is it okay if I go now?" The National Weather Service, the DUAT system, and the FAA briefers exist to bring you the information on which to base an intelligent decision. To make such a decision, you need to know as much as possible about the weather picture, and have a clear idea about what information you can get from the briefer. It's *your* neck that's at stake, not the weather briefer's.

PERSONAL FORECASTING

According to FAR 91.103(a), the pilot in command is required to check weather reports and forecasts for any flight under IFR or not in the vicinity of the airport (whatever *that* is supposed to be). Unfortunately, as general aviation's capabilities have increased, the availability of personalized weather information has decreased, placing the IFR or cross-country pilot at a disadvantage both legally and practically. At many airports the only means of getting a weather briefing is by a long-distance phone call to a faraway flight service station. Assuming a phone is even available, all too often the lines are jammed. Computerized weather, where available, may flood the pilot with information but not tailor it to his or her situation.

A good weather briefing is worth its weight in gold, particularly if (pardon the pun) the situation is clouded. When there isn't a cloud in the sky and winds are light, few pilots will think twice about pushing off on a short flight without checking the weather. But when the ceiling lowers and precipitation develops, everybody wants help in determining the best course of action. It is possible to undertake a flight in the face of less-than-perfect conditions, but only if a good weather briefing is available.

Without such a briefing, you had better stay on the ground, or at least close to an airport. A lack of current weather information limits the pilot's weather knowledge to what he or she can see happening, severely hampering the utility of today's fast, long-range airplanes. Sadly, there's always a pilot or two who is unwilling to accept the stark facts seen through the windshield, weather briefing or not. We read about them all too often in the newspapers. The solution to weather-related accidents is twofold: (1) Learn to be your own weather prophet, and (2) stay within the limitations prescribed by the reliability of your weather data.

WHAT TO WATCH FOR

The best weather information often is not available from the FSS. The pilot who just landed can tell you more than any briefer, assuming he or she is observant enough to recall temperature, cloud height, and drift angles. Get the word from someone who just covered the proposed route and you'll have a better handle on what to expect. Of course, the situation can change in the course of a flight, so temper the pilot's report with your own observations and intuition.

As in the preceding section, get a general synopsis of the weather picture from some source. This information is particularly valuable if you're forced to make up your own forecast. A TV weather map can give you an idea of the positions of highs and lows, as well as the major fronts; from this you judge circulation patterns and possible cloud types. A computer link can give you this wide window, along with specific reports, but you'll still need to custom-tailor the information to your local micrometeorology.

If a low is moving in, expect a general worsening of weather conditions before an improvement starts and a counterclockwise circulation of winds and weather around the low center. If the winds are strong, the low center is probably close by, and if moisture is available to create clouds and saturate the air, flying may be chancy. A nice fat high, on the other hand, will provide a reasonable guarantee of good flying weather, barring such problems as upslope lifting of moisture, which can form low status, or high winds generated by a rapid pressure change.

Fronts may or may not spell trouble, but they must be viewed with at least a cautious eye. If the front is dry, the boundary between air masses might be marked only by a wind shift, a little turbulence, and a few scattered clouds. If there is available moisture being fed to the front by strong winds, the stability of the air mass behind the front becomes highly volatile, generating thunderstorms. The more common cold front is a fast-moving, violent thrust into a warm, moist air mass, with a narrow band or two of thunderstorms associated with the front. If the warm air sector pushes into the cold air mass as a warm front, you can be just as effectively grounded by widespread low stratus and drizzle. Fronts, like low-pressure areas, should be considered a potential threat to your plans, so know the big picture before you attempt to make up your own forecast.

WHAT'S HAPPENING OVERHEAD?

After obtaining an overview of the weather situation take a look at the sky condition, discounting any local pollution. Hazy visibilities obviously mean moisture is present in the air, probably with little temperature decrease at altitude or maybe an inversion (warming with altitude). Haze is not only a hazard to navigation, it is a potential weather maker if a strong push of cold air enters the warm, moist air. In weak weather systems, you may find isolated afternoon thunderstorms sitting in the middle of the haze, even if there is no well-defined front.

An absence of clouds, with blue, haze-free skies, probably means a cold front has passed through recently; if you're heading in the direction of its movement you might delay your departure a few hours to let it move well ahead; then you're probably home free. The presence of a high, thin overcast and some lower clouds could mean that you're on borrowed time; the clear skies are past and another weather system is moving in, probably in 12 hours or less. If you're spending the night, be prepared to make an early departure if you want to run ahead of the weather or a delayed one if you plan to push on after it passes.

Observing weather movement day after day will soon teach you the basic signals: wispy cirrus tells of approaching weather 12 to 24 hours before it arrives, middle-layer clouds start to gather under the cirrus overcast a few hours in advance of the weather, and ground-hugging stratus crowds in as the low center approaches, assuming enough moisture is present. A low stratus deck obscures your view of the sky, of course, and in this case you will have to get an on-top report from an IFR pilot who has penetrated the fog layer. If he says there's nothing but clear sky above the low layer, you will probably have no weather problem after the fog lifts.

Cumulomammatus clouds and rain showers indicate strong turbulence overhead; the forecast obtained in the preflight briefing must be customized for this location by the scene outside.

WINDS AND WEATHER

The velocity of the winds aloft indicates when to expect a change in the weather. If the winds are stronger than suspected, any forecast you have heard had best be accelerated. Observe the movement of low or middle-layer clouds to detect a strong low-level airflow. If only cirrus clouds are visible, note whether they are arranged in streamers, or "mare's tails," because of the strong winds in the upper atmosphere. Wind is weather, as far as pilots are concerned, particularly in mountainous areas; wind always determines the rate of weather change. Wind is movement of air from a high-pressure peak to fill a low-pressure depression. The stronger the wind, the more intense the pressure change will be and the more violent the accompanying weather, depending on the amount of moisture available.

Cloud types affect your flying decisions, of course. Cumulus clouds mean vertical air movement, so you can expect to find turbulence below their bases, forcing you to climb high to seek a smooth ride. A stratus overcast that blocks the sun's heating can assure a good ride at all levels, but if the clouds are thick enough to be a solid layer there may be a low center nearby, collecting the clouds, so you should be alert for lowering ceilings as you fly toward the low. Low scud is definitely a no-go situation unless you're instrument-rated and have enough fuel to make a run for good weather.

How do you judge cloud height? Even the best of us get fooled, but with practice you can learn to gauge a ceiling by watching for such clues as movement in the cloud layer and fragments of clouds. If you can actually see the cloud layer being pushed along rapidly by the winds aloft, the bases probably are very low or the winds are very high. Little or no movement and an inability to detect individual cloud fragments probably means several thousand feet lie between you and the cloud base. A pilot report is the best guide to cloud height unless you want to take a sounding yourself. Beware of "taking a look" if the surface visibility is low; the bases may be only a few hundred feet above the ground, adding moisture to reduce visibility and making even a traffic-pattern flight hazardous. An automated report from a close-by AWOS (automated weather observing system) outlet can help, but even a few miles can make a difference in cloud bases.

TURBULENCE

A pilot needs to know a little about the characteristics of the air mass he or she is observing in order to assess the probability of a smooth flight. Unfortunately, most low-flying VFR pilots prefer to fly in conditions that have the potential for being turbulent. The clear or nearly cloudless blue of the first day following a cold front's passage is advertised as being stable air, but the ride down low is liable to be as rough as a washboard. Below the condensation level, marked by the bases of the puffs of fair weather cumulus, the cold air is being warmed by ground radiation and is reabsorbing moisture deposited by the frontal storms, and thermals will be booming up everywhere. Smooth air at low altitudes, by comparison, will more than likely be stagnate, filled with moisture and pollution, which reduces visibility, a by-product of the air's inability to generate strong vertical currents.

If you take note of the temperature at various altitudes, you'll find that a rapid cooling as you ascend often accompanies clear skies, whereas haze conditions are usually associated with a flat or inverted temperature gradient. If the free-air temperature warms with altitude, then suddenly begins to cool, you'll probably find yourself entering clear air above the haze layer, as the haze ends at the top of the temperature inversion.

The FAA seems to be bent on backing away from providing pilots with current weather information, because as more and more reporting points are being automated as AWOS outlets, FSS functions are limited or outsourced and broadcast weather is limited to just hazardous conditions. Bear in mind that if you use a cellular phone to reach flight service through the 800 WX-BRIEF number you'll be connected to your home FSS, keyed to the area code registered to the cell phone, not the station for the area where you happen to be. Hopefully, there's an accessible phone when you're stuck in Pumpkin Center.

If you're forced to make up your own forecast in an out-of-the-way place, consider the big picture you should always have in your mind, listen to what the locals have to say, check with arriving pilots, and apply your acquired weather wisdom. Above all, don't push your luck beyond your definite knowledge of the weather. The best insurance is still a good weather briefing…if you can just get one.

COPING WITH MARGINAL VFR

What am I doing here?" was my frequent thought as I feverishly tried to compare the limited view of the landscape with my sectional chart. I was a freshly minted commercial pilot, a downy-cheeked 200-hour wonder, back in the days when instrument ratings were for airline pilots. I was off on my first charter trip, slogging my way across the trackless prairie in pouring rain laced with lightning bolts. The paying customer beside me wasn't overly concerned, not realizing the difference between an airplane and a car as far as ease of navigation was concerned. For my part, I was ready to pull over to the side of the road and hole up for a while, but the last airport was a half hour back and I really wasn't all that sure where the next reachable one might be.

As often happens, the god in charge of looking after fools that day permitted the rain to moderate in time for me to orient myself and find the destination, which I thought at the time was due to some pretty neat flying on my part. It was even legal VFR weather, as near as I can remember, because I was quick to duck below 1,200 feet AGL when the visibility dropped below 3 miles; it wasn't that I was concerned with technicalities, I just wanted to stay within sight of the buffalo chips.

Somehow, selling aviation has always bordered on fraud. We tempt a nonpilot with stories of how we made it from Chicago to Nashville in 3 hours, never telling about the 3 days it took to get back home. Mostly, we're inclined to forget about the trips that didn't run on schedule and remember only the uneventful ones that wafted us above the mere earthbound mortals with our seven-league boots. As spouses and business associates soon learn, an airplane is not a flying automobile; a pilot's word is worthless when weather intervenes. The brochures always feature shiny new airplanes winging through clear blue skies, carrying jubilant occupants to always-VFR destinations. Like the 70s song said, I learned the truth at 17—'taint always like that.

Far from abandoning the VFR airplane as a tool of transportation, we need only to face reality and make plans for disruption in our schedule. It is quite possible to fly under less-than-tourist folder weather conditions, providing we don't try to keep *every* appointment and fly *every* trip. Even as an instrument pilot, you aren't going to do that—at least not if you're smart.

RULES FOR SURVIVAL

Rule one is *Stay flexible; if you can't go now, go later.* If you can't get through the pass, go around the hills. The pilot who *has* to get through eventually gets killed trying; the pilot wise enough to wait a few hours might have a relatively easy trip. Timing can be everything. The really low scud may only be down on the deck for a couple of hours, but if those are the 2 hours in which you make your attempt, you'll never know what you missed. Go when you can, not when you feel you must. Most business or family appointments can slide a half day and those that can't require alternate travel plans to cover the weather's vagaries.

Rule two is *Believe only what you see in the windshield.* Sure, you're going to get a weather briefing, for all the good that does. Sometimes the briefer knows less than you do, so use FSS as it was intended—for weather information, not decisions. You make the decisions because it is your tail

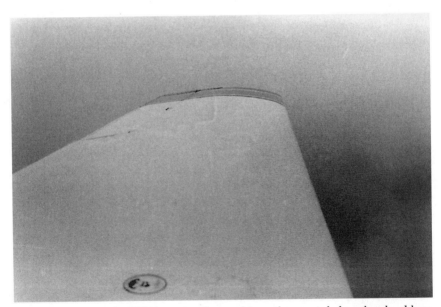

Now *this* is marginal VFR! Dropping down a bit to clearer air below the cloud base may help, but you must set limits on how low you will go. Keep an alternative always in mind!

that gets busted. I had a briefer tell me it was "marginal VFR" on my route. He didn't know I had been checking the weather on DUAT for the past 3 hours, so I asked him where the actual reported weather was "marginal VFR" as forecast. He searched the meteorological reports (METARs), and concluded that one station 250 miles away had 1400 overcast with 7 miles, so the "marginal VFR" was far from widespread and the area forecast was overly pessimistic. On the other hand, a trip the following week was briefed to be a good VFR joyride only to conclude in an instrument approach.

Rule three is *Use what you have available.* A VOR airway is nice to follow in flat country, but not worth a darn in the hills when you have to stay low. A railroad track makes a dandy courseline if it's the only one in the country, but it's a poor one in a metropolitan area. Holding a compass heading in hopes of finding a landmark or a VOR can be a ticket to nowhere if the effect of a strong crosswind is not anticipated.

Rule four is *Never shut the door behind you.* Have an airport at your back at all times, or one in the direction of good weather. I only like to gamble on a sure thing and the weather doesn't fit those odds. Have somewhere to go when Mother Nature fools you. The statement, "Sure was lucky this airport was here," should mean, "This airport was right where I planned for it to be, in case the weather turned out as it did."

DEFINE MARGINAL VFR

What's marginal VFR, anyway? To the FSS specialist, it's a ceiling of 3,000 feet or less and/or visibility of 5 miles or less. To the layman aviator, it's any weather that makes him or her uncomfortable, and sometimes it takes a lot less than the aforementioned FSS parameters. When it's after sundown, that 3,000-and-5 weather is nowhere near our real VFR night-flying requirements, and if we're heading toward rising terrain, even in broad daylight, we want more than 3,000 feet of clearance reported down here in the valley. Your first trip into an area also requires a little better weather than one could comfortably handle after gaining familiarity with the locale. A 30-mile trip creates less sweat than a 300-mile one. We might not even need a 1,000-foot ceiling to find our way home when we're only 10 miles out from the airport.

Now, of course, we're going to avoid flying marginal VFR weather conditions intentionally. But there will be times when the forecast misses or unknown pockets of low weather are encountered. In these cases, your marginal-VFR flying skills had better be up to snuff. Plan your flight with the possibility of coping with rotten weather; be ready to abandon the GPS direct route or VOR airway, relying instead on a major highway, railroad track, or waterway that will take you where you want to go. Stick closer to airports when the weather lowers; these escape sanctuaries will be more numerous along highways, railroads, and rivers. You may never actually have to make a diversion to your low-visibility route, but it's nice to have it planned.

Use all means at your disposal to learn about the weather ahead. As always, know the big picture, so you can turn the right way to find better conditions when your route gets clobbered. Keep up with the latest weather using Flight Watch, FSS, or the broadcast reports available from ATIS or AWOS stations along your route. Ask Unicom stations 25 miles or so ahead how their weather looks; the line person might not be a weather expert but he or she will know if it's raining or fogged in. Pass pilot reports along to FSS as you reach each VOR station to help the chap coming along behind you.

GIVING UP GRACEFULLY

When do you chicken out? Whenever you're having difficulty seeing checkpoints and staying clear of clouds. Stay in firm ground contact, descending immediately or making a 180° turn when you run into scud at or below your flight level. It's amazing how much better the visibility becomes with a few hundred feet of clearance below the cloud base. The legal 500 feet is little enough for good forward visibility. Even if you're IFR rated, don't try to scud-run just to put off seeking a clearance—there could be rocks in those clouds. If you can legally file for an instrument clearance, get onto a route segment that will keep you safely above all obstructions. Don't wander into the clouds in anticipation of a vector.

Set definite limits on how low you will go. An altitude of 1,000 feet AGL is beginning to get right down on the deck, and 3 miles of visibility is about the least one should attempt, even in flat country, with a modern 2-mile-a-minute airplane. It is best to start planning early, looking over the options when you are pressed down to 1,500 feet AGL, and consider a diversion rather strongly at 1,200 feet AGL. If you have a GPS receiver in the airplane, use it to keep track of the nearest airports. Just don't let it tempt you into pressing on because you know right where you are. Definitely execute a 180° turn or deviation to the alternate route when pressed down to 1,000 feet AGL. If you're on a good pilotage route, such as a highway, the temptation is strong to keep going, even down to 500 feet AGL. This is foolhardy; if the weather has deteriorated to that degree already, it will very likely continue to do so. It's time to head for the barn, so don't attempt suicide by continuing on.

A sliding scale of ceiling and visibility values is useful for the VFR pilot. Don't plan on giving up both ceiling and visibility. Three miles of hazy visibility isn't so bad in familiar country with no clouds overhead, but it's pretty chancy under a 1,000-foot overcast. On the other hand, a 1,000-

foot ceiling could be safely negotiated with 25 miles of visibility under-neath and no higher terrain ahead, but even 5 to 7 miles is little enough when the clouds start dropping. Knowledge of the area, the type of terrain, and the availability of alternate routes should influence your decision to continue or land at an alternate.

Above all, don't crowd your capabilities in marginal VFR weather. It's all right to drop down and go a little farther when the clouds thicken up, as long as you have an alternate escape route planned and the firm resolve to call it quits when you reach your preset minimums. Just make that 180° turn before you go too far.

TRAPPED IN
IFR WEATHER

You're a VFR pilot; the license reads Airplane, Single-Engine Land, peri-od. But you've probably had enough instrument time to feel capable of flying in the clouds for brief periods. Think again. As envisioned by the drafters of the FARs, the basic VFR pilot's level of instrument competence is sufficient to make a 180° turn into clear air or negotiate a letdown with-out falling into a graveyard spiral. By no means are VFR-only pilots sup-posed to fly intentionally into a cloud deck.

Suppose, however, that you are heading cross-country and are flying through steadily deteriorating conditions. Because you want to get to your destination, you continue. This leads inevitably to pushing on too far, and sooner or later you'll say to yourself, "How do I get myself out of this mess?"

Start by reasoning out how you got *into* it. Weather doesn't trap pilots; pilots trap themselves by ignoring danger signs and not preparing for possi-ble changes in conditions. You saw the weather develop as you flew toward it, and as a VFR pilot you should be flying in relatively good weather only. Therefore, the way out is the way back. You don't *want* to turn around, but do you want to keep forging ahead badly enough to die trying?

THE HIGH ROAD, AND THE LOW ROAD

Of course, there is always a possibility of getting out of the clouds by going up or down. Exercising this option requires that the decision be made before the troublesome cloud deck solidifies, because you are required to maintain a 2,000-foot cloud separation as you penetrate the layer. Because the breaks will become fewer and smaller, you had best make your move early. Going on top can sometimes seem safe and easy as the beautiful vista of clear blue sky and snow-white clouds lulls one into a sense of secu-rity. One still has to navigate, however, which means relying on position derived from GPS, VOR, or ADF signals unless the breaks in the clouds are numerous; even with a moving map on the GPS screen, VFR pilots will find the regimen of blind navigation to be unnerving. There's also a chance of getting trapped between two cloud layers coming together as one nears a low-pressure center.

Going on top involves a gamble by the VFR pilot that things will open up or at least get no worse as he or she approaches the destination. That gamble doesn't always pay off, resulting in a long on-top retreat with diminishing fuel reserves, looking for a hole. You'll have to get down some-time, and without an instrument rating, equipment, and currency, you must do it VFR.

Dropping down to go under is a good option if you want to maintain ground contact, but you may encounter reduced visibility under the cloud deck and you'll probably be subjected to more turbulence. You'll have to abide by definite limits on how low you can go, depending on terrain and obstructions ahead. Navigation by pilotage is easy, and the GPS works even down on the deck, but radio reception for VOR tracking is compromised by the low flight level. However, it's easier to reach an airport in a hurry by staying under the clouds, one positive aspect of taking the low road.

But you've now tried too hard, and you're really in it. You can't see the ground except for an occasional glimpse, and this cloud doesn't seem to be "just another little puff" as you'd hoped. You've gotten yourself into a

Clouds, clouds everywhere and not a hole in sight. A VFR pilot in this situation had better start working on an escape plan. If an instrument letdown is needed, be sure you're equal to the task.

bad situation, and now you've got to get yourself out. Avoid panic at all costs; grabbing the mike and hollering for help won't bring somebody to take over the controls. Fly the airplane first, and in due course you can ask for assistance.

AS IF YOUR LIFE DEPENDED ON IT

Concentrate on keeping the wings level, using the top index of the artificial horizon. Relax the death grip on the control yoke, or you'll first climb, then overreact with a dive. The nose of the airplane belongs on the horizon bar, but remember it'll tend to go there by itself if you'll just relax and let the trim tab do its job. Now that you know how to stay upright, look at the upside-down number on the bottom of the directional gyro and begin a turn; roll out when you reach that heading. Then, ignore the directional gyro for a minute or so, because you won't need it, and concentrate instead on flying the artificial horizon to stay in a 15° to 20° bank with the nose level.

How do you do a 180 if you have one of the old horizontal-face DGs? Well, you can add 200 to your present heading and subtract 20, which turns 115° into 295°, or take away 200 and add 20 if your heading is already past 200°. If that's too much to handle, just hold the bank at 15° to 20° for 1 minute, then roll out, and you will be close to a 180° change in direction in the average lightplane.

After completing the 180° turn, hold the wings level and stay on your chosen heading. You are now on the way to the better weather you left behind, and a few minutes of straight and level flight will probably bring improved conditions. If it doesn't, you can begin a cautious letdown, as long as you are reasonably aware of your location and can estimate a safe level-off altitude, so you won't penetrate a mountain peak. Should you be unsure of your position in a mountainous area, you'll have no choice but to climb until reaching an altitude that will clear the terrain. Then you may as well confess your emergency and get help to find your way down.

If a descent is called for, reduce power by 300 to 400 rpm in airplanes with a fixed-pitch propeller, or 4 to 5 inches of manifold pressure if you have a constant-speed prop. Allow the nose to drop below the horizon until the attitude gyro's horizon bar rests on the top of the artificial airplane's nose dot or gull wings. This should yield about a 500-fpm descent rate at cruise speed, with no trim change required; keep the wings level, don't apply back pressure, and wait for the ground to appear. Try to form some idea of how low you can go and still have 500 feet or so of terrain clearance; be ready to abandon the descent in case you're in an area of zero-zero fog. Because you're heading back toward better weather, you'll most likely break out after a few hundred feet of descent. Remember to level off at least 500 feet below the cloud base, so you'll not only be legal but will benefit from improved visibility as you remove yourself from the moisture.

WHEN ALL ELSE FAILS

Now for the worst possibility: You've turned around, descended, given up, and climbed, all without seeing anything but cloud. Survival now depends on keeping a clear head and confessing your predicament. I realize the standard recommendations for VFR pilots trapped in weather are to climb immediately and ask for assistance, but making a 180° turn and letting down can often eliminate the need for ATC help, if action is taken promptly. Never forget who has to fly the airplane; if you seek ground assistance you'll still be the one who has to hold a heading and stay upright.

But if you don't see any other way out, get help to find the way down. While maintaining a heading, turn up the volume and open the squelch on the radio. Dial in an FSS frequency and attempt to contact flight service. They won't have ATC radar, but they can refer you to someone who does. Just give them your call sign, general location, and frequency on which you're listening. If you're requesting a reply on the VOR frequency, don't forget to turn up both volume knobs, nav as well as comm. After you get a reply, confess your situation and ask for assistance in getting back to better weather.

If you don't get a reply, you may need to climb higher to establish line-of-sight contact with the ground station's antenna. If all attempts at communication fail, your best course of action is to track inbound toward a navigation station along your general route of flight and try again later. Your preflight preparations should have included forming a mental picture of where blue sky lay so you would know which way to head if things got really bad. However, you probably omitted this step or you wouldn't be in this mess now.

If the situation is urgent, go directly to 121.5, the emergency frequency, which could save a few minutes while FSS looks up a center frequency for you. You can, of course, call up center yourself if you have the frequency available. Be prepared to follow instructions and answer questions when you establish communications with an ATC controller. He'll want to know your aircraft type, number of persons on board, and hours of fuel remaining. You'll also be asked if you're instrument qualified and equipped; you'll have to answer in the negative, so you'll be asked if you want to declare an emergency, unless the controller wishes to assume the obvious—that a VFR pilot in the clouds *is* an emergency. You may as well declare; if you had any other choice, you wouldn't be calling ATC.

FOR EMERGENCY USE ONLY

When an emergency situation has been declared, the rulebook can be discarded and you can get priority handling to conclude your emergency as quickly as possible. Expect identification procedures, in the form of a transponder code or identifying turn, so you'll be easier to see on radar. If you used 7700, the transponder code for emergency, you'll probably be pulled off that code after communications are established.

If radar identification is made, if you can fly the airplane well enough to hold a heading, and if you're not low on fuel, it should be simple for ATC to direct you to an area with higher ceilings and away from instrument traffic, where you can let down into the clear and land with no more emergencies. The controller would prefer to keep you in firm radar contact, so he or she will probably try to hold you above a certain altitude for tracking purposes until you're near the field. Don't argue about the choice of airports; cooperate by getting on the ground and ending the emergency for everyone.

In case you are having difficulty handling the airplane and keeping up with the controller's barrage of instructions and questions, don't be afraid to speak up and tell him or her you're having trouble with the airplane. Remember, you are receiving assistance from the ATC system, but it can't fly the airplane for you.

After your encounter with IFR weather is over, whether it involved a contact with ATC or just a few seconds of sweaty palms while you escaped the fog, take a cold, hard look at how you got there so you won't let it happen again. Doing something dumb can serve a purpose if you are honest enough to tell yourself, "Hey, that was stupid." Be sure you find a way to avoid making the same mistake twice. Review your weather-briefing technique, and get with an instructor to talk it over. You may not get another chance to mend your ways.

SURVIVING
THE SEASONS

WIND WISDOM

It was a day fit for only the hardy and the half-witted. The wind sock snapped to and fro, trees sighed as their branches bent with the wind's force, and airplanes rocked nervously in their tiedowns. Pilots who succeeded in landing their aircraft offloaded ashen-faced passengers and told of encountering "severe turbulence" aloft. Clearly, it was no day to go flying...or, was it?

Perhaps a rough, windy day isn't the most comfortable time to be flying, but it can provide a chance to improve your flying skills. If you want to grow as a pilot, you must treat everything as an opportunity, including some rough air now and then. The number of accidents resulting from loss of directional control during takeoff and landing leads us to believe that pilots aren't staying sharp on their runway handling, probably because more and more airplanes are being used for transportation, an application that provides few takeoffs and landings to maintain proficiency.

So, if you want to stay sharp, take advantage of this gusty afternoon and fly a few circuits in preparation for the day when you get caught in just such conditions while coming in with a load of passengers. It isn't possible to avoid all encounters with strong winds, and if you pass up this chance to practice you'll be just that much more rusty when the inevitable showdown comes.

HOW MUCH WIND IS TOO MUCH?

Two limitations are placed on our operations: that of the airplane and that of the pilot. No matter how well-designed the airplane, some recently licensed or rusty pilots are just not qualified to fly in winds of more than 15 knots, at least not until they gain more experience by flying in progressively stronger winds. On the other hand, even the best pilot can't handle a wind that exceeds the airplane's ability to provide control. Such control limits are usually reached during crosswind operations, either on the runway or while taxiing. For a rough rule of thumb, figure on running into difficulty whenever the wind speed is greater than one-half the airplane's stalling speed, so if your bird has a 60-knot V_{so}, operating in winds of 30 knots will be stretching the airplane's limits, even if the wind sock is pointing straight down the runway.

Another axiom that has kept me out of trouble on many an occasion states, "If you can't taxi, don't fly." When the wind is so strong that you have difficulty holding the airplane straight while taxiing, or if the gusts tend to lift a wheel off the ground, forget it—put 'er back in the shed and wait for a better day. The use of brakes to control one's taxi path is not unusual in tailwheel aircraft, but most nosewheel-equipped airplanes shouldn't be operated in winds strong enough to require brakes for directional control. The exceptions are those planes with a free-castering nosegear, such as the Grumman, Cirrus, or Diamond singles.

Even if you observe those two limitations, there will be plenty of days when the winds are strong enough to make you wish you were somewhere else. However, you won't learn anything by leaving the airplane in the tiedowns and flying it only on perfectly calm days. Practice, even if only for a half hour or so, hones your skills for that day when the winds pick up while you are en route. There are pilots and there are passengers; make sure the person sitting in the left front seat of your airplane is a *pilot*, ready to handle any and all flight conditions.

PLAN FOR THE WORST

Undertaking a bit of windy-day practice requires some forethought, such as checking the forecasts to see how strong the winds are supposed to get (maybe they haven't peaked out yet). Other runways should be available, in case the wind shifts or the currently acceptable crosswind component increases. The alternate runways need not be on the departure airport, but they should be nearby, well within range of the fuel supply.

How do you determine the wind's speed? If a tower, FSS, or Unicom is on the field, you can always ask the controller or radio operator for a wind check, but if a ground observer's report is not available, other methods will do. Check for a report from a nearby ATIS or AWOS/ASOS (automated surface observing system) outlet. Finally, look at the wind sock. A normal wind sock will generally fill to a straight-out position with approximately 25 knots of wind; drooped to a 45° angle with the bag still filled, the sock will signify winds of 10 to 15 knots. When the sock is bent the winds are down below 10 knots. Twenty-five knots are also indicated if trees are tossing their branches about briskly; if entire trees are swaying before the gale, you are probably experiencing a 35-knot wind. If you are within sight of a lake or river, 15 knots of wind will generate a few whitecaps, and 25 knots will produce whitecaps over most of the surface, probably with wavetops blowing spray downwind. At 35 knots, windstreaks of foam will be seen moving across the water.

The crosswind is not going to upset this pilot, who is making sure he maintains runway alignment while touching down on his upwind wheel.

Taxi slowly as you move out to go practice; the greater the wind, the slower the taxi speed needed. Use minimum power and brakes during taxi so the airplane will not lean downwind on sharp corners. Hold your controls properly to keep the wings and tail from being lifted by the wind; if taxiing upwind, hold the stick slightly aft of neutral (full aft for tailwheels), and if a crosswind component exists, hold the ailerons full into the wind. Taxiing downwind, the stick goes full forward with ailerons held away from the crosswind. If you can't remember control positions, just look at the ailerons and elevators and imagine the position that will keep the wind from "getting under" the airplane and lifting it. In this way, you can see the benefit of keeping the elevators and the windward aileron deflection down in a tailwind, directing the wind over the airplane rather than under it.

Assuming you arrive at the end of the runway without undue difficulty, park with the airplane headed into the wind to avoid damage from control surfaces being slammed to the stops, as well as for improved engine cooling and better stability when you increase power for the runup.

THE WINDY TRAFFIC PATTERN

After the checkout, line up carefully for the takeoff. If there is a crosswind component, begin the takeoff run with full aileron held into the wind.

Maintain nosewheel contact for steering during the takeoff, but do so without putting excessive loads on the nosegear. Accelerate to a speed that will assure an immediate liftoff into a climb, then rotate the nose and "jump" into the air. Be prepared for a short takeoff roll—the airspeed indicator needle will move off the peg almost as soon as the throttle is opened.

Once off the ground and climbing, it's just like any other day, except for the turbulence buffeting the airplane. Unless there is visible evidence of an end to convective currents in the form of flat-bottomed puffs of cumulus, don't expect to be able to climb above the rough air. When surface winds reach 30 knots, moderate turbulence may extend to over 7,000 feet AGL.

Someone will have to land this bucking, bouncing airplane, so there's no point in postponing the inevitable. Let not your heart be troubled, as the saying goes; **just have a plan to keep from bending the airplane.** Have your throttle hand at the ready to provide either a blast of power to arrest a sudden sink or full power to initiate a go-around. One can always go around when the situation becomes unsalvageable, and knowing that you have such an out will prevent undue concern from interfering with your flying. Don't fret about the wind, just go around if you can't handle it and give it another try. After several unsuccessful attempts to land, proceed elsewhere and land on a runway more directly aligned into the wind or at an airport with fewer burble-generating obstructions. Even a landing in a pasture is better than creaming the airplane, if it offers an into-the-wind touchdown.

By arriving at the touchdown at just the right moment, you can sometimes catch a lull in the wind and effect a fairly normal landing even after successive go-arounds, another reason to make several attempts. It is well to land using as little flap as the runway length will permit. Some airplanes have a stalling speed several knots higher with the flaps up and the increased speed at touchdown can mean better control will be available through the landing. When stall speed is nearly the same with flaps up or down, find out which configuration gives best control and visibility in the touchdown attitude.

The touchdown should be made on the main gear with the nosewheel well clear of the runway. There is no need to try for a full-stall landing on a windy day; sacrificing control by holding off too long can lead to trouble. The best technique is to place the airplane in a slightly nose-high attitude a foot or so above the pavement, then just hold the attitude as ground contact is made. After the airplane settles on, continue flying the controls through the rollout by applying additional aileron into the wind and maintaining runway alignment with rudder. Keeping the nosegear in light but positive contact with the surface aids steering. Counter heavy brake application with up elevator to keep the main wheels in contact; some airplanes require immediate flap retraction to improve control. Relax only when you've got the ropes tied to the wings, as the old-timers say.

A windy day is no cause for panic, although the pilot's workload is certainly increased by the challenge of the gusts. Make it a point to go out and do battle with the wind occasionally, just so you will be prepared to do some serious flying when the sock is straight out.

THUNDERSTORMS: SUMMER MONSTERS

Practically everyone knows that you just don't mix airplanes with thunderstorms. Failing to stay clear of these anvil-headed monsters has brought many an airman to grief, leaving his widow to make the best of things. The stock advice is not to take off on a cross-country flight when the route forecast includes the likelihood of thunderstorms.

Unfortunately, this advice is too all-encompassing to be realistic. Before long, a low-time pilot (known as the 200-hour menace in insurance circles) will observe other pilots coming in from a route he has just canceled out, and they will tell him it was an easy trip, just bumpy with scattered showers. After a few of these experiences, there will come a time when he decides to take off anyway, in the face of a T-storm forecast, and he too makes it through without incident. Henceforth, he will discount any and all warnings about possible thunderstorm activity, and will continue to press his luck until he either gets the living daylights scared out of him or makes the papers.

The fact of the matter is that there aren't too many long trips that can be flown in the summer months without encountering an area where conditions are right for thunderstorm formation. To keep you aware of this, the weather service will probably issue a forecast that includes, "Chance of a few scattered afternoon thunderstorms, occasionally becoming severe or forming into broken lines." The storms may not occur, or may not affect your particular flight, but the chance of you successfully making the trip depends on many factors, some of which are not under your control.

First, **as in any weather situation, know the big picture** so that you can deviate slightly to avoid the worst activity. Next, try to find out the type of condition spawning the cumulonimbus cloud (Cb) development; are the storms popping up at random in an unstable air mass, or are they associated with a frontal system? Seek information about where the storms are currently located, if any, and which way they are moving.

Deciding whether or not to attempt the flight in the face of a "possible thunderstorm" forecast also depends on the difficulty of the route and the type of options available. If there are few airports along the way, if unfamiliar or rugged terrain prevails, or if other weather considerations are already chancy (poor visibility, ridges obscured), you had best not try threading your way among the miles-high storms. Certainly if the route forecast reads in such stronger terms as "moderate to severe thunderstorms likely, forming in broken to solid lines," your time would be better spent seeing to the snugness of your tiedowns.

On the other hand, if the flight is not too long, is over easy country, good options exist, and the route is otherwise uncontaminated by weather, a cautious attempt might be possible. If you can see no towering cloud tops in the distance, in haze-free air, you ought to be good for 50 miles of exploration. A haze layer necessitates a climb to get on top of the smog before a good visual assessment can be made. Departure timing is all-important; a route open for one pilot can be completely blocked to someone coming through an hour later.

At this point, some discussion of thunderstorm theory might be in order. The cumulonimbus cloud grows in stages, first as a small popcorn-shaped cumulus, then rapidly gaining height and size until rain begins to fall. Dissipation occurs as the rain continues, the entire cycle elapsing in a span of 1 to 3 hours. Other storms develop during this time, of course, making the life cycle of a single storm rather academic to the stranded aviator.

Lifting of unstable, moisture-laden air is the act that triggers the thunderstorm formation. This lifting can be started by thermal heating, by an

This building cumulus cloud is already above our altitude and will soon be a thunderstorm. Circumnavigation is the only safe option.

overriding of warm, moist air over a cold layer, or by air being pushed up the slope of a mountain range. Whatever the initial cause of lifting, under the proper conditions the warm, moist air continues to rise on its own, building skyward past the condensation level where clouds form. The rising current of air continues to carry more warm air up, building the tops higher and higher.

Eventually the moisture-laden air reaches saturation, raindrops form and begin to fall, and the Cu-type cloud changes to a mature Cb. Typically, the building cumulus stage contains updrafts and the mature storm carries both strong updrafts and downdrafts in close horizontal proximity. This severe wind shear can overstress an airplane, if the pilot does not lose control in the chaos of the storm and tear it apart himself. Naturally, thunderstorms are the home of other goodies such as hail, heavy icing, lightning, and even tornadoes—no wonder we are told to stay clear!

HOW TO HANDLE THEM IN FLIGHT

How clear should you stay? From a single, moderate storm, perhaps 5 miles would be adequate, while a big severe storm deserves a 10-mile berth. Beware the clear sky beneath the overhanging anvil, way up there; hail often falls through this area, as the storm leans over in the upper-atmosphere winds. The point of the anvil indicates the direction of the storm's movement, if you can see it. Middle-layer clouds and building cumulus can obscure your view of the big Cb, lurking in the darkening haze ahead.

If a decision must be made about flying IFR, don't. An instrument clearance is of scant help when coping with thunderstorm avoidance, unless you have the performance to get on top of the lower cumulus and have a cooperative controller. Without radar on board as well as on the ground, your best option would be to stay at an altitude below the cloud bases, where you can see the rain areas falling from the biggies, and can hunt an airport quickly if the situation becomes untenable.

Air mass thunderstorms are frequently isolated and well scattered, and pose little problem other than an increase in fuel burn because of circumnavigation. An early start helps as the instability of the whole witches' brew increases as the sun begins to beat down. Your best trick is to be headed away from the unstable air into another air mass type, or be tied down before midafternoon.

FRONTAL AND OROGRAPHIC STORMS

Frontal storms can be troublesome, depending on one's direction of flight. They are generally bigger, meaner, and have a greater tendency to hold hands when aligned with a cold front. If your route parallels the front, you may be a witness to one of nature's most spectacular displays, without becoming a participant, as you fly several hundred miles uneventfully. But you won't be able to reach a destination 20 miles away, if it happens to lie on the other side of the storm line. When tempted to break through a small gap in a line of storms, remember that more than one line of storms may precede the actual front, and you may be in worse trouble if you enter that devil's doorway to be encircled by storms. You want 5 miles of distance between you and the storms on each side, and a clear view of an unlimited horizon on the other side, before penetrating the line. Cold frontal storms generally move rapidly along the front itself, from south to north, while the entire system drifts slowly eastward.

Orographic, or upslope, storms form against the windward side of a mountain range, making them a real problem if they take root in the only passes available near your route. If you are approaching from downwind of the ridgeline, you may cross it to find a valley full of violent storms, so be ready to retreat. Do not attempt to fly under the cloud bases, because upslope storms frequently shroud the higher terrain of the peak in rain and stratus.

The Great Plains of the country's midsection spawn some of the most widespread thunderstorm activity, providing an unobstructed flat region

From a distance, these air mass storms are beautiful, but they are getting too close together for a comfortable flight through their area. The anvil head of the right-hand storm is typical of a mature Cb and indicates the direction of movement. The vertical lines at the top of the picture are window reflections.

where cool Canadian and Pacific air masses can crash headlong into damp Gulf air. The results can be lines of storms stretching across several states, building up to 60,000 feet, with wind gusts to 100 mph in squall lines, and an occasional tornado. If you have a chance to see this sort of weather coming, you aren't likely to tackle it.

REVIEWING THE BASIC RULES

Basic rules for flying in or near thunderstorm areas include staying in VFR conditions and not flying into any rain showers. Even torrential rains seldom affect the engine's performance, but poor visibility spells instrument conditions and a possible loss of control in turbulence. More important, you have no idea what lies beyond the obscuring rain showers, and entering that wall of water might take you into a trap of encircling thunderstorms. Don't accept just a peek of sunlight through a gap in the rain shafts—it could be the sun shining down a "chimney" in the middle of a solid mass of cells. You want a good horizon line with visible ground features as assurance of open skies beyond, and even then remember to keep miles between you and the storms as you fly through a gap. Narrow openings in a line have a way of closing up as you near or enter them, and once through it would be nice to be able to back out, should the prospects beyond the gap not be attractive.

For in-flight help, ask Flight Watch, reachable on frequency 122.0 almost anywhere in the country above 5,000 feet, for updated radar reports, which give you an idea of the size, coverage, and movement of major precipitation areas. With this information, at least you will not head blindly into a worse situation. The Flight Watch frequency can be a convenient listening post for information given to other pilots, even if you choose not to call.

Above all, never proceed into thunderstorm areas without a clear avenue of escape to an airport, because conditions can worsen rapidly. Keeping one eye on what's going on behind you or in the direction of your "out" is vital to avoid the unpleasant and possibly fatal task of penetrating a thunderstorm in an attempt to escape your entrapment.

Flying around thunderstorms might often be necessary to effectively use an airplane in the summertime. Just keep on top of one of the fastest-changing situations in weather, and always maintain an "out" so you can land to sit out a big line of cells, rather than tackle them in the air.

SQUALL LINE ENCOUNTER

In all of nature's weather arsenal, there is perhaps no more awesome sight than the approach of a rapidly moving line of thunderstorms. The clear, slightly hazy sky will contain a few distant thunderheads by midafternoon, easily dismissed as isolated air mass storms that will live out their lives in their present location and be of no great hindrance to cross-country flying. Should there be a strong weather system moving into the region, however, these distant storms may draw closer together and intensify, developing as a squall line ahead of the cold front.

You will usually be warned of this possibility if you secure a weather briefing from a flight service station. Even if the weather check is skipped, however, or if the FSS briefing is running a little behind the actual situation, it takes no great amount of weather wisdom to tell what's going to happen. A portion of the sky that has grown visibly darker in the past hour or so must have something pretty potent behind it, and no one can look up at the towering side of a rapidly building cumulus without feeling rather insignificant and small.

WATCHING THE GIANTS APPROACH

As experienced from the ground, an expectant hush falls as the line approaches. Not a leaf stirs. The hot, muggy air leaves the wind sock hanging limp. The darkening mass of water-filled storm cloud stretches to 40,000 feet or more, blocking out the sun's light as it nears. A few puffs of stratocumulus may float past, torn loose from the front of the storm line like bits of cotton candy. The viewer hears an artillerylike rumble of thunder and sees abundant cloud-to-ground lightning. Dust swirls up from open fields well in advance of the rain, the first gusts of the storm hitting the ground like the bow wave of a giant ocean liner. A cold upper-atmosphere wind strikes suddenly, spinning the wind sock around from its listless parked position with 40-knot gusts.

Minutes later, a roll of low stratus cloud sweeps by a few hundred feet overhead, its condensation caused by the sudden introduction of cold air to the warm, moisture-laden air at ground level. As pilots frantically run to check their tiedowns and control locks, the rain hits, first with large, pellet-size drops, then blowing in with hurricanelike sheets as the storm intensifies. Wind gusts of 60 knots or more are commonplace.

Fortunately, the tempest's fury is short-lived; thunderstorm lines are usually no more than 20 to 40 miles wide, and within 30 minutes the rain will slacken and the winds will subside as the storm moves on, leaving overturned airplanes and sodden turf in its wake. It's possible for tornadoes to be lurking within these fast-moving lines, but the twisters rarely are seen and most damage reports actually originate with the strong straight winds. Heavy hail can occur in localized areas, and light hail will be found in the clear air outside the rain shafts, falling from the overhanging anvil downwind from the storm. Severe wind shear can be encountered near the ground as an aircraft climbs or descends through one layer of moving air into another.

AS SEEN FROM THE AIR

Impressive as the storm appears from the ground, it is even more awesome when seen at close range in the air. Flying through turbulent but sunny skies, the pilot awakens to the fact that his windshield is growing dark. A glance upward through the haze reveals giant towers of cumulus ahead, and the wise pilot will check for a suitable landing spot close by. Pressing on, as many will, discloses the lightning-filled maw of the squall line. Heavy rain fills the void from the low cloud base to the ground, and the cloud stretches up to infinity. The air may be relatively smooth until within a few miles of the actual storm cloud, but when the first gust of the storm line is felt, there will seldom be further need to convince the pilot to go elsewhere.

Thunderstorm turbulence is of a different quality than that experienced in normal flatland thermals, more on the order of a strong mountain wave. The airplane is no longer capable of being precisely controlled, and the pilot feels as if she were in the grip of a giant, malevolent hand that is tossing the airplane up and down like a toy. These are the giant swells of the air, breaking into heavy seas where opposing currents meet. The airplane is simply carried away, rising smoothly with the vertical speed indicator (VSI) needle pegged at 2,000 fpm or dropped wrenching with the seat belt cutting into one's lap. The pilot's best course is to turn tail and run at the first contact with a thunderstorm's vertical currents, before penetrating any heavy rain that will block a clear view of retreat. If this option has already passed and the aircraft is committed to a penetration of the storm line, a power reduction should be made to slow the airplane to maneuvering speed and concentration applied to maintaining a constant heading that will hopefully lead one out of the storm after a few terror-filled minutes. Thunderstorms are like alcoholism: Abstinence is the only sure cure.

CLOSE ENCOUNTER OF THE WORST KIND

My most memorable brush with a fast-moving thunderstorm line came not on a cross-country flight but at a prosaic airport picnic, one of those informal affairs where everybody brings a little something and much good fellowship is exchanged. There being no aerobatic aces on hand, I was enjoined to add a little color to the proceedings by flying a few maneuvers in a Cessna 150 for the crowd. Nothing illegal, to be sure, just the usual training stuff that might be demonstrated in a commercial pilot program— 60° steep turns, an accelerated stall and a two-turn spin, with a few lazy eights and some chandelles thrown in to gain height for the finale, a deadstick landing with a rollout calculated to end in front of the hangar. Strictly show-off stuff, heavy on the grandstanding and short on precision—or, in this case, common sense.

A squall line advances across the countryside, bringing a wall of water and intense lightning. The low cloud warns of cold air pushed ahead of the line, and the car headlights on the highway tell of poor visibility even for driving. The winds went from calm to 40 knots-plus a few minutes after the picture was taken.

As I taxied out, I realized that the line building in the northwest for the past hour or so had grown closer and I wondered about the wisdom of putting on my act ("It'll only take 10 minutes; surely the rain will hold off that long"). I made a max-performance V_x climbout with all the pizzazz a 150 is capable of developing, and climbed toward the storm line to check it out ("If it's closer and meaner than I think it is, I'll be in a position to run for the barn"). I could see no nearby rain shafts, the dark cloud base was well above me at 3,000 feet AGL, and I launched into my few minutes of tomfoolery for the benefit of those watching below. I noticed a strong tendency to drift away from the storm. This stiff horizontal flow should have warned me that things were moving fast.

As soon as my brief sequence was finished, I chopped the mixture back and pulled up to stop the prop, noticing an approaching roll of gray cumulus. The storm was coming fast; I needed to be on the ground *now* and it would take 5 minutes to execute the deadstick approach. I checked my altitude and was shocked to find it *increasing* rather than decreasing! I pushed over instinctively, a reflex action to avoid the threatening wall of dark cloud at my elbow; I was indicating well over 100 mph and the VSI showed a 500-fpm climb with the prop stopped. There was no turbulence, just a smooth, elevatorlike upward pull. The situation was out of control; I could see the cloud base nearing rapidly. I was about to go IMC (instrument meteorological conditions) with inoperative gyros, sucked up by a huge vacuum cleaner.

To heck with the airshow! I turned to a right angle from the storm line's axis and ran away from the airport, heading for the clear air at V_a speed, trying not to exceed maneuvering speed in case I hit rough air that could overstress the airplane. I cranked the engine back to life to regain suction power for the gyros and to get a little extra drag from the windmilling prop. At the same time, I lowered 40° of flaps for maximum drag. Lowering the flaps reduced the protection offered by maneuvering speed, but I was wanting out of this and would try anything to increase my sink rate. After several long, anxious seconds of running from the storm the VSI dropped below zero into the "down" half of the dial, and I ran the flaps up and began to breathe a little easier.

ANY PORT IN A STORM

It was time to review the situation. There was no weather ahead; an alternate airport lay some 25 miles away. On the other hand, the home field was only a few miles behind me. A glance over my shoulder showed that the automatic airport lighting system had turned itself on because of the darkness overhead, but there was no rain and I could see under the storm for 10 miles or more. I decided there was time enough to make one try at the runway. I expected heavy winds at the surface and, if I had to abandon my landing attempt, I could head for the alternate. I bored toward the runway with cruise power, making good a ground speed of perhaps 60 mph. The final approach was flown at 90 mph with flaps up, to make any sort of headway, and I only reduced power as the threshold neared. Fortunately, the wind was directly down the runway, because the sock was standing straight out and whipping, a sign of 35 knots plus. The touchdown was as soft as a thistledown due to the ultraslow ground speed, and several pilots wingwalked me into the lee of a hangar to receive the welcome due a victorious warrior returning from battle. Minutes later the wind was 90° to the runway with heavy rain.

Instead, I should have received the reward of a bonehead, with a severe flogging for endangering both an airplane and my own skin. Mistake number one was taking off in the face of oncoming weather simply because of the press of "business," plus continued adherence to the planned activity until it nearly became too late. Only the god who looks after fools gave me a dry cloud base and winds aligned with the runway just long enough to make it home. It would have served me right to have spent the night sleeping in the airplane at an alternate field, fighting mosquitoes in the heat.

MANAGING
HEAT STRESS

When the heat's on, it brings a host of problems for the summertime aviator. If you live in a part of the country where the midday temperatures crowd the century mark, or if you visit such areas, be ready to deal with the temperature's trials.

As you no doubt know, airplanes in general, and low-wing airplanes in particular, get beastly hot inside the cabin when sitting out in the sun. With the doors and windows tightly shut, it can get well over 130° inside, enough to warp plastic plotters and make a pilot's thinking processes less than sharp.

The heat's on! Midsummer flying brings discomfort and stress for both pilot and airplane, but it can be done if you plan for it.

So, do what you can to cool things down inside. Park in the shade if at all possible, open the doors and windows for ventilation while doing the preflight inspection, and keep the time on the ground to an absolute minimum. Grass parking areas tend to be a little cooler than the sun-baked expanses of concrete or asphalt. Invest in a sunscreen of space blanket material to cover your windows when parked, or fabricate something similar on your own.

TAKING CARE OF THE ENGINE

Don't forget about the torrid conditions under the cowling, where the engine lives. A modern air-cooled power plant was designed to be operated in flight, where the cooling air can blast around the fins and carry the heat out of the cowling. As the power output demanded from engines has been steadily increased through the years, cowlings and baffling have been refined for the ultimate reduction in drag to best utilize this extra power—all of which leaves an engine that can't be operated for prolonged periods on the ground, or in high-angle climbs, without overheating.

Therefore, be as easy as possible on the engine; don't warm it up too long, just enough to stabilize the oil pressure before taxiing. By the time you reach the end of the runway and complete a runup in 100°F weather, it'll be ready to go. Face into the wind if parking for any length of time, double-check that the cowl flaps are open, if you have them, and monitor the temperature gauges. The cylinder head temperature will react faster than oil temperature, but remember the instruments are only measuring the health of the engine at one specific point, so there's no guarantee that another area of the engine isn't already overheating by the time a redline is indicated.

PREVENTIVE MAINTENANCE

Before summer arrives, take the cowling off and put a vacuum cleaner to work sucking out the leftover bird nests from between the cylinder fins and baffles. This accumulation can cause hot spots and maybe a top overhaul. The seasonal oil change would be a logical time to clean out the engine compartment, as well as to check for loose or damaged baffle plates.

The engine's oil level should be kept topped off without overfilling, because this is the major heat-exchanging medium within an air-cooled engine. If the oil is allowed to get a few quarts low, it just isn't going to carry away as much heat as it passes around the engine.

Hot weather also plays havoc with radio gear. It's a good idea to check all the cooling provisions—such as vent openings, fans, and blast tubes—

that keep the black boxes operating at their best. Don't run every radio in the house unnecessarily while on the ground unless your cooling setup is able to handle it. It's true that today's solid-state units don't generate as much heat as the avionics of 50 years ago, but their displays frequently put out copious warmth, and the inner works are not particularly tolerant of high temperatures; consequently, they must be protected from potential heat stroke.

HOT STARTS

Getting the engine restarted after a fast fuel stop can often be a trial, especially for fuel-injected airplanes. Everyone is hot and miserable, anxious to be back upstairs in the cool, while the engine cranks fruitlessly. The problem—and hence the solution—varies from aircraft to aircraft. It can be vapor lock, the particular plague of fuel injection; a fuel line gets warm enough under the confines of the cowling to cause a vapor bubble to form and block fuel flow. The answer is to force the vapor through the lines with positive fuel pressure from the boost pump, or perhaps to start the engine on primer if available. This carries the danger of overpriming, perhaps accumulating excess fuel in the induction system where it can catch fire as the engine starts. Use the boost pump only until fuel pressure is steady, then shut it off to attempt a start. Don't use the hand primer without cranking, and if you do get a stack fire, keep cranking to suck the flames into the engine, where they can do no harm.

The opposite extreme is flooding the engine with excessive fuel so that it fires only fitfully with puffs of black smoke. Just pull the mixture into idle-cutoff and crank once more, pushing the mixture back into full rich when the engine comes to life; opening the throttle speeds the process. Some engines flood easily when hot; others won't start without plenty of fuel: It's best to try a lean start first, then add fuel later if that doesn't work. Best of all, just follow the instructions in the pilot's operating handbook; the people who built the airplane knew best.

SAGGING PERFORMANCE

Summer brings a compromise in overall aircraft performance; the old bird lifted off readily with a full load in the winter, but it's a different airplane when the mercury soars. The low density of the hot, humid air saps power from the engine, and also hampers the wing's ability to produce lift, adding up to our old friend, density altitude trouble. So what if the field elevation is only 3,500 feet? Crank the temperature up to 100°, maybe more just above the surface of the runway, and it becomes 7,000 feet for all the airplane knows. Pad those figures in the owner's manual plenty, and don't think your troubles are over just because you made it into the air. The rate of climb is going to be less than outstanding, perhaps only a couple of hundred feet per minute at full bore, and strong downdrafts will be more than you can outclimb despite your best efforts. A few minutes of fighting your way uphill at V_y speed and your engine will be hollering for mercy as the oil temp nears the redline.

The moral of the story is: **Don't fight the temperature; learn to live with it.** Fly early and late, avoiding the full heat of the day whenever possible. Don't load up your airplane to the limit; leave a seat empty instead. Climb at higher speeds to keep the engine cool, and level off momentarily if the temperature gauges leave the green. Don't yank the nose up to fight a downdraft, just keep going and fly out of it; your flight path may be erratic, but the engine will stay cooler.

The usual summer flight plan is to climb until you find smooth air, then level off at the next available cruising altitude. This often puts you on top of the scattered cumulus puffs that mark the rising thermals. Make sure you don't get pushed higher and higher by the building of baby thunderstorms until you are suddenly in the middle of one that is outclimbing you. Better get back down below if the tops threaten to engulf you. The rough ride at lower altitudes can generally be avoided by flying early in the day.

Don't overlook the effects of summer on you, the pilot. Heat and humidity are tiring stresses, and a fever is a fever, whether it comes from an illness or an overheated cockpit. Keep your body full of liquids and take a rest break now and again, letting your copilot have a turn at the wheel. Use a set of noise-suppressing headphones so you won't irritate the passengers by battling the shrieking ventilators with a loud loudspeaker. These are little things, but accidents are often the result of little things that add up. When you keep your cool the flight just seems to go better.

Summer is a great time to fly. Enjoy the fair skies and long daylight hours, but respect the special terms of this hot-weather season.

HAZY SUMMER DAYS

Low-visibility haze conditions are increasingly becoming a fact of life for cross-country pilots, especially during the warm-weather months. Flying in the semi-VFR whiteout of severe haze can be an unnerving experience for low-time pilots, and a nuisance even for experienced ones. There are few tasks more difficult than keeping one's aircraft straight and level when the only ground reference is straight below. Taken to extremes, this difficulty winds up under the accident classification "attempted continued VFR flight into IFR weather conditions."

HAZE DEFINED

Haze is a restriction to visibility with no measurable cloud present. It differs from fog in that a fog layer can be seen and measured; the cloud is at the surface or slightly above it, with a relatively low top. Fog with a surface visibility greater than five-eighths of a mile is reported as mist, or BR in the international format. (The French word for mist is *brume*.) Haze, on the other hand, is not a visible cloud; the atmosphere is simply filled with water vapor and pollutants that scatter the sunlight. It can be tenuous and difficult to measure, and it often extends to 10,000 feet above the surface. It can be a local condition caused by heavy pollution and moist air being "capped" over a city by a temperature inversion, or it can be a widespread air mass condition, when moist air stagnates instead of following the general west-to-east movement across the continent. Wooded hills are often wrapped in the blue haze of warm moist air restricted in its movement by ridges and valleys.

COPING IN FLIGHT

It is by no means simple to avoid flying in marginal-VFR haze. The ground visibility can be quite acceptable, with perhaps 5 miles reported, whereas the in-flight visibility can be a mile or less, particularly near the condensation level where small puffs of cumulus are attempting to form. You can depart with every assurance that the en route weather is good VFR, and find yourself in practically IFR conditions. It behooves every pilot to be ready to cope with such an eventuality.

If you are instrument rated and prepared to file IFR if the need arises, you will probably opt for flying in VFR conditions on top of the haze, which can require a cruise altitude of 8,500 to 12,500 feet. This will place you in clear air, where you can see approaching traffic and distant weather in unlimited visibility. This is also a tempting choice for the VFR pilot, but one that should be used with discretion. Flight above a haze layer can be the same as flying over a cloud layer; the top may not be horizontal, giving you the "leans" because of a false horizon. You will be totally dependent on GPS navigation or VOR fixes, and to avoid getting lost or entangled in special-use airspace (SUA) you should stick to the airways or routes that you know will keep you clear of SUA. Because you must descend VFR, be alert for a ceiling developing below your flight level.

On short flights it might be impractical to climb above the haze, and it will be necessary to fly down in the soup with everyone else. IFR pilots will feel more comfortable with a clearance, but they should be aware that VFR traffic is quite legally boring around in the haze with them, and they should keep a sharp lookout. If you are flying a fast airplane in a busy traffic area IFR, with no autopilot, press a front seat passenger into service to help keep watch.

On top of the haze layer, this summertime pilot must navigate carefully with the limited visibility when flying down in the murk.

The VFR pilot must respect his or her limitations in haze, and those of the equipment. It is possible to fly in somewhat worse conditions with a GPS moving map and an artificial horizon and gyro compass than with a basic 1946 instrument panel, but even down in Class G airspace you shouldn't continue indefinitely with less than 3 miles of visibility, or in any case with 1 mile or less. There are too many unmarked towers out there, and the faster the airplane, the less time you will have for avoidance.

NAVIGATION CHORES

Pilotage navigation will be difficult at best as the landmarks loom up and disappear before you have time to check them fully on the chart. It is not wise to draw a line across the countryside and expect to follow it as you might in clear weather. Low-level winds aloft reports being what they are, you can easily drift off course by 5 miles in short order, and that's enough to leave you disoriented in 3-mile haze. It is far better to use an alternate route that offers good visual references, even with a GPS tracking your flight. One of the best methods is to swallow your pride and lock onto a good highway, river, or railroad track, following it to your destination. By doing so, you will know part of your location at all times by having a line of position, and it will be simpler to identify each town or checkpoint as it comes down the road. Airports are generally located along these lines of surface transportation, so you can easily make that 180° turn and follow the road back to a safe haven if it gets too bad.

When forced to follow such an impromptu course line as a road, make note of the time over various points along the way, writing the minutes figure on the chart beside the point. Should you become disoriented later, you will at least have a basis for knowing your last position and the time over it.

The time of day naturally has much to do with the in-flight visibility in haze. An early morning eastbound trip, or an evening westbound flight, can reduce 6 miles of reported surface visibility to less than a mile at altitude. The light rays are scattered and diffused by the haze to create an almost total whiteout when looking into the sun. Try a lower altitude in such cases to bring more of the ground into view for attitude reference. Should you be flying with the sun at your back, remember that approaching traffic will be blinded by the upsun viewing angle, and may not spot you until dangerously close.

VFR IN IFR CONDITIONS

The greatest challenge to flying VFR in haze is the continual switching from visual to instrument references and back again. You will be looking at the ground below, comparing it to the chart, and then, upon glancing back at the panel you will notice that your heading has changed 10° or 20°, or that the altitude has slipped up or down a bit. You make a correction, probably by using the artificial horizon and other instruments, then go back to looking at the milk bottle void outside, once again drifting away from straight and level flight. Learn to relax at the controls while looking outside, flying "loose" just as you would in good VFR conditions. Panic and fighting the controls will not help your situation.

Fortunately, the country is still fairly replete with VOR stations, and you can often keep a running cross-check of your progress by dialing a "from" reading from a station to confirm your exact position when landmarks slip by without ample time to check them out. Global positioning can make us too bold for our own good as we wander through the haze with our eyes fixed on the GPS screen instead of outside looking for towers and traffic. You can still get a DF (direction-finding) steer from an FSS that advertises this service over certain outlets. Being too proud to admit you're not sure of your position has taken many a pilot past the point of no return.

OTHER WEATHER HAZARDS

Be alert for an embedded thunderstorm lurking in the haze like a shark in the surf. If the haze suddenly darkens ahead, take notice; a boomer is probably casting its shadow on the top of the haze layer. If the cloud is only a good healthy building cumulus, with plenty of clear space underneath, lighter skies will reappear beyond the cloud base. But a spatter of rain on the windshield should mean a 180 for the VFR pilot, back to the last airport. Dodging embedded cells is a game for those equipped with airborne radar or sferics (lightning detection) devices, and even then it's pretty exciting.

Should you be running into scud, losing sight of the ground occasionally, don't press your luck. Stay below the clouds, even though you have reports of excellent weather ahead, if they keep lowering as the haze thickens; don't gamble on sneaking through with only a few hundred feet of altitude and no visibility. You are probably trying to bust through a stationary front, and the good weather reports are from the cold air sector beyond the down-to-the-ground frontal zone. Many good pilots have died trying to make it those last few miles, so don't add your name to the list by flying into a hillside or tower.

The term *air stagnation* has become all too familiar as mankind continues to foul its own nest and nature occasionally pours a strong temperature inversion into the witches' cauldron. Pilots will find themselves forced to fly in haze or not make use of their airplanes at all, so you had best learn to cope with it, within your limitations.

FOG, INSIDIOUS ENEMY

Fog is one of aviation's undefeated evil triumvirates, its partners being thunderstorms and icing. Although fog may not bring airplanes swiftly to earth when first encountered, its ability to eventually halt all operations is certain. To cope with fog, we've developed Category II instrument approaches, but few general aviation pilots and airplanes are approved to use them. However, the CAT II approach permits only landings down to a visibility of roughly one-quarter mile and a decision height of 100 feet; beyond that we get into the even-more-stringent standards of Category III operations. The final transition to the ground must still be made with visual reference to the runway unless one wishes to trust the triple computers of an autoland system.

There are times when the fog thickens even below instrument landing minimums. Grounded, despite all our ratings and equipment, we pace the floors of airport lounges and spend hours on the phone reshuffling plans, cursing our fate, and wondering when the weather service people will learn to cast their chicken bones properly so we can get an accurate forecast. It's important to understand the nature of any enemy if you intend to challenge it successfully. Fog is an enemy, to be sure.

Forecasting fog with any degree of precision is guesswork. It's easy for a prognosticator to foresee the chance of fog occurring, but that doesn't mean it will necessarily happen. A degree or two of temperature can make all the difference in the world, as can a few knots of wind or a change in the amount of moisture in the air. Fog can be so localized that forecasting from any distance is impossible. You must be your own forecaster, or suffer the consequences of a useless forecast.

WHEN WILL FOG APPEAR?

Fog is basically a cloud base at the surface, obscuring visibility rather than forming a ceiling overhead. A condition reported as mist or haze is simply fog of a different degree, when saturation of the air is not complete. When fog is at the surface, a typical observed ceiling will be reported as "100 obscured" or "vertical visibility 100 feet," meaning that the ceilometer penetrates 100 feet into the cloud, not that the base is 100 feet above the surface as with a normal ceiling report. Looking forward through the windshield from an altitude of 100 feet requires perhaps five times as much distance to see the runway as looking straight down; obviously, you aren't going to see anything but fog when attempting to land. A report of a partial obscuration may tempt you to try to get in, but have an alternate handy because "partial" can become "total" on the very next ATIS.

The thickness of a fog layer greatly affects your chances of getting into—or out of—a particular airport. Although fog typically burns off after sunup at the rate of about 100 feet per hour, a cirrus layer holding back the sun's warming rays can keep it from breaking open at all, and added moisture from an overrunning layer of air will refuel the fog generator. Still, just being able to take a look at the actual sky condition above the fog can make all the difference.

IFR LAUNCH DECISIONS

There probably isn't an instrument-rated pilot in the world who can resist a glimpse of clear blue sky through a break in a fog layer, even if the runway

The weather is far below IFR minimums in heavy fog. The only safe approach is to wait for an improving trend to show itself.

visual range is only 1,200 feet. That revealing hole is all he or she needs to load up and go, assuming there is an alternate field within reasonable range to handle contingencies after departure. The pilot knows there will only be about 1 minute of actual instrument flying on each end of the trip. On the other hand, should the sky have that thick, dark, gray look characteristic of solid multilayered weather, the trip will be much more challenging, requiring greater care in selecting alternates and balancing fuel reserves, particularly if the traffic flow is heavy enough to create delays in receiving approach clearance. A recent pilot report is worth more than all the guesses of the weather service.

FOG FORMATION

Whence cometh the fog? On its little cat feet? From condensation, my dears, only that and nothing more. Water vapor condenses into visible cloud when the air mass reaches saturation, and one need only add more water to an air mass of a given temperature, or reduce the temperature of an air mass with a given percentage of saturation, to trigger fog. No moisture, no fog; it's that simple. This explains why temperature and dewpoint can often sit perilously close to each other without fog appearing, or why fog can appear when the spread isn't razor-thin.

If the moisture is present already, nighttime cooling can reduce air temperatures to the condensation point, generating patchy radiation fog. A strong flow of warm, moist air over cold, frozen ground can create widespread zero-zero conditions as advection fog. A weak cold front can stall out against a warm air mass, fogging in the frontal boundary. Or a sea breeze can move cool offshore moisture tightly against a wall of steeply rising terrain to sock in the coast. These are all examples of fog's origins.

Is fog strictly an instrument pilot's problem? Hardly, considering that clear air is normally only a few hundred feet away vertically, inviting the VFR pilot to fly over and take a look, or to depart in hopes of finding a hole. These sucker traps are deadly, and require cautious planning to avoid.

Our purpose here is to give you advice on how to cope with fog, rather than to provide a meteorology lesson, but it is important for you to understand the basics so you'll know when you can't win. There's no point in trying to get into an airport that's never going to get up to minimums, and you shouldn't bother dragging your passengers out to the airport just to sit around the waiting room. Take all fog predictions with a barrel of salt, realizing that forecasters can't foresee every local aberration in terrain, wind speed, or upper cloud deck; remember, the weather is what it is, not what it's forecast to be.

BETTING ON YOUR CHANCES

When faced with a cheery forecast but a poor current METAR, assume the skeptical stance of a used-car prospect. Don't bet your life on any forecast until it starts to happen. You may be legal to launch without an alternate according to the forecast, but you'll wish you had one when the "clearing after seventeen-hundred local" doesn't materialize. A burnoff is possible only if certain conditions are met—no more moisture coming in, no higher clouds, no change in wind estimates. When the forecast doesn't start to come true by midmorning after a night of fog formation, throw it away; you can bet that an amended one will be coming out at the appointed time.

Another classic sucker job is the predawn departure under the stars, a trap that boxes in many VFR pilots. Anxious to get an early start, our hero dutifully checks his route while half-asleep, finds everything clear, and launches with his landing lights ablaze. There's a hint of trouble when condensation trails are seen wisping from the prop tips on takeoff. The sun peeps over the horizon an hour or so later, and minithermals start mixing the encapsulated pockets of moisture near the ground. Nonhomogeneous streams of low-lying fog in the valleys start to climb up out of their beds, swallowing the plains, and suddenly there is a blanket of white everywhere. Hopefully, a few hilltop airports will remain open to receive visitors, or maybe the sun will work its magic before the fuel runs out. Meanwhile, one more pilot has received his education about fog.

The cure for fog is avoidance, just as with our other two assassins, thunderstorms and ice. Avoidance means adapting a forecast to your own location's conditions, asking a lot of questions of the natives if you're not at home, and tempering your boldness with an honest appraisal of your ability to fly. Lots of fuel and a wide-open VFR alternate—these are the tools we rely on to defeat the insidious enemy called fog.

COLD, COLD START

It comes at the end of every autumn. It's that time of the year when the engine oil turns to Jell-O, frost coats the windshield, and last summer's battery loses its ability to spin the prop rapidly. Winter is back and the annual battle to start a reluctant machine is joined once more.

Getting the bird started is an elementary task in warm temperatures; just pump the throttle once, twist the key, and you're off and running. But when the temperature drops below 40°F the task is no longer so simple. Pilots of southern-based airplanes may find this out the hard way when they make an overnight cross-country into the northern states. There, stiff oil and tropical starting procedures just don't cut it. They will all too often wind up summoning aid to get the airplane started—unnecessarily at times. By taking a few minutes to think about cold-weather starting, you may save yourself embarrassment and a few bucks.

WHAT'S THE PROBLEM?

It's no mere accident that airplane engines are hard to start when cold. They are practically designed to be that way, with large per-cylinder displacement, heavy-viscosity lubricating oil, and a relatively crude updraft carburetor, further hampered by a battery smaller than that used by a garden tractor. Compared with automobile engines, it's a wonder the darned things start at all. Sometimes they just *won't* out of sheer cussedness—usually when a hurried departure is necessary to meet a schedule.

Tricks of the trade exist, however. First, preheat the engine if at all possible; that's old advice, but it's still sound. Bringing the cylinder heads, oil pan, and induction system up to 50°F or so ensures a prompt start, without washing the residual film of oil from the cylinder walls by heavy priming. The good folks in the northern border states know all about this and they can't believe that pilots in the middle layer of the country will actually attempt starts in near-zero weather without preheating.

ADDING A LITTLE HEAT

Preheating apparatus need not be elaborate. If a small heat source is used, such as a heat lamp or a firepot, it will be necessary to drape a tent around the cowling to contain the heat. Take care to keep flammable materials a good distance away from the heat source, and stay nearby to watch over things during the warm-up. Electric dipstick heaters can be useful to thaw the stiff oil, and by turning the prop over you can pump warm oil through the engine, but you should be aware that some of the oil-filler tubes are plastic, and consequently should not be subjected to heat. The fastest and most reliable heat is delivered by an oil-fired space heater, the kind that looks and sounds like a small jet engine, either blowing its heat through flexible ducts or simply aimed into the cowl openings from the back of a pickup truck. Fifteen minutes of the dragon's breath and ol' Betsy will start at the touch of a key. The best preheating setup is a permanently installed system of electrical heating elements on the engine, almost standard equipment in the north country but seldom seen on southern visitors.

Nearly all preheating requires electricity, and that can be a limiting factor. The tie-down or hangar might be far away from an electrical outlet, and the footing too slippery to allow walking the airplane to the source of

A jury-rigged preheating setup, usable with electricity near the airplane. The space heater must be kept far enough away from the cowling to avoid scorching paint, and equal time is given to both cowl openings. The windshield requires deicing also.

power. If there is no other choice, a cold start may be attempted, but it is not only hard on the equipment, the outcome is far less sure. Down to about 20°F a cold start is generally possible, given some experience with the airplane's characteristics. Below that figure, chances are reduced to 50/50 at 10°F, and about nil at 0°F. There it will take luck, skill, and several degrees of incantations to bring the bird to life.

Most of our starting-procedure tips are directed toward carburetor-equipped engines. Fuel-injected airplanes have an easier time of it but are still hampered by the poor performance of the battery and the thick oil. The advantage of fuel injection stems from the availability of positive fuel flow directly to the cylinders; carburetor-equipped airplanes will have priming systems that can deliver fuel to the intake manifold while cranking, but it often takes more hands than you've got to prime, crank, and work the throttle. All cylinders are not primed evenly, if they are even primed at all; many airplanes have systems that prime only one or two cylinders.

GIVING IT A TRY

Begin by verifying that the ignition switch is turned off and by expending a little of your own muscle, carefully turning the propeller by hand to limber up the oil so the battery won't have to do it. Then prime the engine several strokes, leaving the primer pulled out, and immediately turn the propeller over by hand about four compression strokes to load the cylinders with fuel with the throttle wide open. Jump into the saddle, pull the throttle back, and hit the starter before the primer fuel evaporates. Make sure all lights and other electrical items are off so every available amp is directed to the starter motor.

If the engine fires, keep it running by using the primer or the carburetor's accelerator pump, by rapidly pumping the throttle through the first half of its travel. Avoid pumping the throttle with fuel-injected engines, however, because it has no effect other than to confuse the fuel-metering mechanism. Just hit the boost pump to keep a fuel-injected engine going if it tries to die.

IF AT FIRST YOU DON'T SUCCEED

If you don't meet with success immediately, don't keep cranking, expecting a miracle. If there's no response after a dozen revolutions, save the battery and try something different. Look under the cowling for dripping fuel, which is a sign of excessive priming and throttle pumping. This causes the gasoline to drain back down through the intake pipes and out the carburetor body, pooling in or under the cowling. As a fire hazard, this is one of the many undesirable features of a cold start. It pays to have a fire extinguisher handy in case a backfire through the intake pipes ignites the pooled gasoline. Don't panic if a fire is detected in the carburetor, just pull the mixture to idle cutoff and continue cranking the engine to draw the flames into the intake manifold where they can do no harm. Move the airplane away from the spilled gasoline, if possible, or wait for it to evaporate.

When dripping fuel is seen, do not prime further but open the throttle fully and crank the engine to clear the flooded mixture. Be ready to pull the throttle back if the engine starts (to prevent high rpm, which can cause extrahigh oil pressure against the bearing seals). Most likely there will be no start, and it is best to discontinue cranking after a few revolutions. Opening the throttle and turning the prop by hand with the switch off to save the battery can also clear the engine. Then close the throttle and begin all over again.

If no dripping primer fuel is seen, add a few more priming strokes and engage the starter once more. It helps to have the primer charged and ready for action in case it's needed to keep the engine running, but push the primer knob in slowly so the engine won't flood. Fitful attempts to run are usually a sign of flooding, whereas no firing at all is more common and an indication of insufficient fuel. Immediate starting followed by silence can be taken to mean primer fuel was burned but the induction system fuel was not reaching the cylinders in time; try again, adding more priming fuel.

PROPER ENGINE CARE

By all means, **keep the engine running after it starts;** if moisture is present in the air, a few seconds of firing followed by a shutdown will invite condensation in the form of frost on the spark plug electrodes and there can usually be no further ignition until the plugs are removed and dried or the engine is heated. Watch the tach carefully for the first 60 seconds and pump the throttle or hit the boost pump if it falters.

Battery life is usually measured in seconds during winter starts, particularly if the battery has several seasons on it. Jumper starts are often necessary and can provide the extra amps sorely needed by the starter motor. Just watch your polarity when hooking up to avoid cooking the alternator, and make sure all avionics are off to keep them safe from odd voltages coming through the system. Avoid having extra people around during jump-start operations so a preoccupied onlooker won't walk into a prop while helping with cables or whatever. If you have one of the newer airplanes with a 24-volt battery, forget about jump-starting from your car; you'll have to hunt for an FBO with a battery cart.

Fuel vaporization is difficult with high-density cold air, even after the engine is running; warming up with the carburetor heat on may be advisable to help the engine accelerate when throttle is applied. In extreme cold, this may actually induce carb ice by warming the carburetor into the ice-prone range, but such occurrences seem to be rare. Keep some power on during descending flight to maintain heat in the engine and avoid shock-cooling the cylinder heads by rapid heating and cooling, which can cause cracked cylinders. Handle the throttle gingerly, taking two seconds to reach full travel to avoid stalling the engine.

Take advantage of any winterization equipment provided, such as baffles for the cowl openings, insulation for the oil tank and breather pipe, and bypass routing for the oil radiator. These items, like preheating, will add hours of life to your engine; each cold start probably costs the equivalent of 2 or 3 hours of wear, so you should avoid needless starts and short trips. If you *must* go when the mercury is near the bottom of the tube, give it your best and perhaps you'll get away with it.

CHILLER ICE

Ice! The very word seems to have a chilling effect, as when shouted by a bowsprit lookout on a sailing schooner groping its way through a fog-shrouded North Atlantic. Soon after pilots began trying to fly on through the winter months, they ran afoul of icing's insidious effect on airplanes, and the battle between the pilot and the elements has raged ever since.

Learning conditions that foster ice formation, and the options available to cope with it, is time well spent for all winter-flying pilots. For most of the lighter segments of aviation, the best remedy will be avoidance. Don't go where the ice is, and leave it where you find it. When you play the game of how-much-will-it-carry, particularly in a minimally equipped airplane, you're playing with fire, not just ice, and every winter several pilots get burned.

These words are not just for the cloud fliers; a note of caution is also in order about VFR icing. Oh, yes, ice can mess up your VFR flying in several ways. No doubt we were all introduced to the carburetor heat control early in our careers; some of us may never have had cause to use it in anger, yet Ol' Man Carb Ice is out there, waiting, along with another VFR ice trap, freezing drizzle.

CARB ICE

Remember, aircraft carburetors are, for the most part, updraft installations, wherein the carburetor hangs below the engine in relatively cold air. When a liquid expands into a gaseous state, such as when fuel is vaporized in the carburetor, it quite naturally gets cool, just as you cool off by evaporating perspiration from your body. Thus, in the absence of rising warm air from a hot engine block, temperatures in the aircraft carburetor throat can drop below 32°F in an ambient temperature of 50°F or more, and if moisture is present in the air, ice will form on such projections as the carburetor's venturi. This has a choking effect because of the restriction of air volume but not fuel flow, and it causes the engine to lose power because of the excessively rich mixture. The condition is terminal in extreme cases, and if you

141

wait until the engine's fire has gone out you will no longer have a source of heat to deice the carburetor. Early detection and prevention is the answer.

At the onset of carb ice, a slight power loss is noticed—a drop in rpm if flying a fixed-pitch propeller, or a loss of manifold pressure with a constant-speed prop. Thinking the throttle has crept back, the pilot will nudge the power back up. Soon, power will be back down again. If she doesn't get the message, she'll shortly wind up at full throttle with power still dropping off. For warning, check the prevailing conditions—look at the outside air temperature. If the temp is down in the Fahrenheit 40s or less, and there's any moisture around, you've certainly got carb ice. Moisture can be present in the clear air just below a cloud layer, in hazy but legal VFR conditions after fog dissipates, or downwind from large bodies of open water. Naturally, precipitation such as light rain or moist snow can do an excellent job of putting ice in your carburetor.

For fast, fast, fast relief, pull the carb heat knob full on, and leave it there despite the unnerving consequences. A badly iced-up engine may quit running briefly, bucking and snorting as the melted ice passes through its system. If you keep the faith, power will suddenly come back with a rush, and all your excess throttle will make its presence felt. You may, after clearing the carburetor throat, have to adjust the mixture for smooth operation because of the lower density of the hot air, but leave the heat on

Alcohol anti-icing, as installed on this airplane, can be used to prevent ice from sticking to the propeller blades; channels in the rubber boots at the blade's cuff stream the fluid out along the metal blade. At best, any lightplane ice protection merely buys additional time to escape.

regardless. A carburetor temperature gauge or ice detector is a nice insurance policy, if you have a spot on the panel for one.

There is a type of induction icing that can also plague noncarbureted airplanes in the form of air filters blocked with moist snow. Alternate air intake sources should restore power if you encounter this problem with your fuel-injected engine.

VFR AIRFRAME ICE

Ice on the windshield can readily shake up a supposedly VFR pilot. If you don't have a good defroster on your airplane, fly only in the clearest winter air. If you do have one, make sure you're familiar with its operation before winter sets in. If you should get into icing conditions, you at least want to be able to see to get back on the ground. Shutting off the cabin heat outlets will direct extra hot air to the windshield in some airplanes.

The VFR pilot can pick up a rapid glazing of clear ice in freezing rain conditions, despite relatively good visibility. The light rain showers look normal enough, inviting a direct flight through them, but if the outside temperature is below freezing, watch out! The drops hit with more than the usual solidity, and within seconds your leading edges are painted a new shade of white. You've just had VFR ice, clear of clouds. Avoid any further showers and, if conditions are widespread, land at the first available aerodrome.

ICE AND THE INSTRUMENT PILOT

Instrument pilots, of course, know the patent remedy for freezing rain. Because it is falling from a warm layer of air overlying the cold air in which you're flying, you go up into the warm layer to get relief. Do not do so blindly, however, expecting no further trouble. Keep fuel and alternatives available in case you penetrate the upward slope of the cold air mass again, or if you run out of climb ability before reaching the warm air.

Flying in clouds at temperatures below freezing would seem to be a perfect setup for icing. However, IFR pilots learn that ice is difficult to forecast accurately, because they make flights unscathed when expecting to find trouble and pick up ice on other flights when no mention of it was made in the forecasts. Ice can be a very localized phenomenon, and the freezing level is not always level, often having ups and downs that cause you to run in and out of ice formation. In general, unstable air masses with vertical air movement will make ice formation difficult to predict. Even pilot reports may not be reliable, because one aircraft gets only frost and another reports moderate to heavy ice.

In any event, ice accumulation is your signal to take action. Rather than passively watching it build, hoping you'll get there before it gets too heavy, make some plans. Check weather reports for airports with above-freezing surface temperatures, in case you need to divert, and warn ATC you've got some ice and may want to change altitude. If ice continues to accumulate, *do* something. Request another altitude—either higher, hoping to get on top of the clouds or between layers, or lower if the surface temps are well above freezing. Don't forget you can always turn around and go back to where the ice wasn't forming—*you're* flying the airplane, you know, it's not being pulled to the destination on rails.

Even with anti-icing gear on your airplane, make contingency plans when you first hit ice. Light aircraft boots and prop deicers are a compromise in weight versus capability, and it is unwise to think in terms of being able to handle any weather you come across. Anti-icing equipment is for buying time, allowing you more options to extricate yourself from the icy trap. Keep the wings and props waxed or siliconed if you will, but remember the Arab adage: "Trust in Allah but tie your camel."

Should you be forced to land with an accumulation of ice on the airplane, beware the increase in stall speed and drag caused by the blunt form and uneven surface of the ice. Tack on several extra knots and carry additional power for the approach, but not to the point of coming in too hot to land on the available runway. Use minimal or no flaps, because lowering the flaps can cause an increase in the tail's negative angle of attack and make it stall when it's already operating at its limit with a coating of ice.

Some of the best flying of the year comes in the winter, when the air is smooth and the density altitude is low. Ice is just a fact of life we must accept along with winter's benefits. Understand it and take action when you encounter it; don't wait until all your escape routes are closed.

FLYING SNOW

When winter spreads its annual blanket of white over much of the United States, pilots who fly after snowstorms are often faced with a clear sky but an uncleared runway. Operators of north country airfields usually make every effort to plow the runways and ramps as soon as possible, because another layer may come with the next weather system and the snow would soon be too deep to handle. In the lower tiers of snow states, however, where the white stuff comes less often, smaller airfields may be dependent on hard-pressed road and street crews to clear the city's runways. The airport can remain unplowed for days while the trucks fight to keep the roads open. Right or wrong, that's the way it happens, and the wise pilot always calls ahead to check the condition of the field he or she intends to use after a snowstorm.

CHECKING IT OUT FROM THE GROUND

Assuming you've encountered a thoroughly snowed-upon airport, how would you cope with it? If you are on the ground, the obvious first step will be to survey the depth of the snow cover, either on foot or from a suitable vehicle. Using a car with street tires is asking for a long, cold walk when you get stuck in a drift at the far end of the runway, so a four-wheel drive vehicle, tractor, or snowmobile would be best. In addition to reconnaissance, the vehicle's tires or treads will leave tracks useful for reference during takeoff and landing on fresh snow.

How much snow is *too* much? As with grass, mud, or sand, it depends on the airplane involved. A two-place airplane with fat tires can negotiate 4 or perhaps even 6 inches of loose snow, whereas some four-placers might be stopped by 3 inches. Larger aircraft sometimes can make up for a lack of flotation with brute horsepower, but it's not wise to make a practice of taxiing with nearly full power. Tightly cowled engines can become overheated, because of a lack of adequate airflow, without the pilot's knowledge, especially when headed downwind.

Packed, dense snow on a hard runway presents far less challenge than loose, drifted snow, such as might be found on a turf runway. Fortunately, airports are generally built in the open countryside where the wind blows freely, and you may find your problem is getting through the drifts around the hangar, with the runway itself having accumulated only an inch of hard snow. Asphalt runways melt snow nicely once the sun gets through the white reflective layer of snow; the black absorbs heat rays and gains enough warmth to thaw even at subfreezing temperatures.

AIRBORNE SITUATION ASSESSMENT

If you are airborne when you encounter a snowy runway, you are obviously in a much poorer position to make a decision about using the strip. Unicom can help, but not every airport has radio equipment, and flight service knows only about the status of the airports whose operators choose to call in. If you see a large "X" marking the center of the airport or at runway ends, take the hint and go elsewhere; the field or runway is officially closed.

Lacking any information about the field, you can try dragging it at an altitude of 50 feet or so to look for clues to the snow depth. Runway light fixtures, markers, fence posts, parked airplanes, and so forth can help gauge the accumulation, but these should be backed up by watching for tire tracks left by other traffic. Two continuous tracks probably mean only cars have used the runway; three light ruts vanishing into unbroken snow cover reveal aircraft usage, and the more tracks there are on the snow surface, the better you can feel about your chances for a safe landing.

If you see no sign of activity, however, you had better not use the field. There must be a reason for the absence of clues and signs of life. It may be a foot of powder snow! Attempting to land on a fresh snow surface is like playing Russian roulette with a derringer; the stuff may be only a couple of inches deep, or it may be a couple of feet. Take your chance and see what you get, a smooth rollout or an inverted airplane. Aside from uncertainty over the accumulation, depth perception ranges from slim to none over fresh snow. You could fly right into the ground on your approach, so land on unbroken snow only in an emergency.

AIRPLANE CARE

The airplane requires careful preparation for a postsnowstorm flight, above and beyond the normal preflight inspection. If possible, remove the wheel fairings so they cannot accumulate snow and frozen slush. When the aircraft has been parked outside, plan to spend considerable time digging it out. If the first snow was melting as it fell, you may have a lift-destroying

This large metropolitan airport is open, with all runways plowed, but some ramps are still ice covered. Suburban fields may not be so lucky, however.

layer of rough ice on the wing surfaces, which must be removed before flight. A heated hangar, an alcohol spray, push brooms, and time to spare may all be required to cope with the snow-covered airplane.

Pay particular attention to openings in the aircraft where blowing snow may have entered. Uncovered cowlings can pack full, not only through the front openings but through the lower outlets where the cooling air exits. Watch for snow accumulated in the carburetor heat inlet, where it can melt in flight to give your engine a diet of water. The air cleaner is invariably obstructed by packed snow. Remove loose snow from the insides of the control surfaces, and dig any obstructions out of the control surface gaps. Aircraft with stabilator tails require careful deicing to keep the balance of the flying tail unchanged. In-flight flutter can develop with an out-of-balance stabilator. Stabilators also require large openings in the aft fuselage, through which snow can enter in quantity. Make sure all drain holes are open, so a day of thawing temperatures won't leave you with a 50-pound block of ice in the tailcone on the morning after.

TAXI AND TAKEOFF TECHNIQUE

With the runway checked and the aircraft prepared, you are ready to do battle. During taxi, expect poor braking action on packed snow and more than adequate drag in areas of loose snow. Use full-up elevator when the airplane is moving to keep the nosegear light, or to keep the tailwheel down, depending on your type of landing gear. Avoid starting and stopping in deep loose snow, which can be picked up by the propeller and plug an air filter. Run up on a clear spot, if available, where the brakes can have a chance to hold.

The takeoff should be accomplished in the best soft-field style, using the recommended flap setting and lifting the nosewheel or tailwheel as soon as possible. It is important to keep the tail low in any case, just clear of the snow surface, both to gain maximum lift from the wings for an early liftoff and to minimize the tendency to nose over if a patch of deeper snow is encountered. Plan to use twice as much runway as normal, and choose a point at which you will discontinue the takeoff attempt if acceleration is poor. If a sluggish takeoff is aborted, you can make another attempt more successful by holding to your previous tracks to pack a widened path when taxiing back, which you can use for increased acceleration next time.

After liftoff, enjoy the view and relax. Retractable gear aircraft should have the gear operated through a few cycles to make sure all snow and sludge are dislodged before finally stowing the gear in the wells. There is much less clearance than most pilots realize, and numerous malfunctions have resulted when snow and ice have packed into wheel wells. Be prepared to change your navigation habits over a snow-covered landscape, because many familiar landmarks are altered or missing. A black roadway

If the snow starts falling again, this airport may soon be closed once more. Drifts are forming downwind of old snow piles, blocking nearly half the pavement.

can be hidden beneath a snow blanket, town roofs blend into the land-scape, section lines may be indistinct, and you will find yourself calling upon old compass skills, the GPS map notwithstanding.

SNOW LANDING

The landing, like the takeoff, is performed at the slowest possible speed, which normally means full flaps and stick all the way back for a full-stall touchdown from a foot or so above the runway. The rollout will be short if any appreciable snow is on the runway, so leave the stick full back after touchdown and rely on aerodynamic braking. Tailwheel airplanes should be plopped on tailwheel first if control permits, giving an anchor effect to help hold the tail down as the main wheels touch the snow. If used care-fully, a little throttle can hang the aircraft in the air for an even slower touchdown. Don't be surprised if one or both main wheels are locked up because the brake puck froze to a wet disk after liftoff, particularly if you still have the wheel fairings installed. Fear not, just steer aggressively, using equal brake to offset the locked one, and wait for a spot of dry pavement to appear, where the tire will roll and break the ice loose.

Okay, you're out of gas or being crowded by deteriorating weather and you think you must land at an unused airfield with unknown snow depth. Try to define the runway limits; if runway lights are visible, remember there is normally 10 feet between the lights and the pavement's edge. Judgment of flare height will be a problem, and the lights can be of further help in this case. Other useful aids are bushes along the runway, boundary fences, and the like. If you have any dark, disposable items on board you could make a pass down the runway and drop them in the touchdown zone to give you some depth perception as you land. Have your passengers pad and brace themselves in case the snow is deeper than you anticipated. Stall out just above the snow and hope for the best.

Winter flying can be beautiful, but don't be the first to land at a strange field. You may have to stay longer than you'd planned!

USING YOUR HEAD

GO/NO-GO
JUDGMENT

Recognizing the limitations of one's abilities and those of the airplane one flies is basic to the continued survival of any pilot. Yet, all too often we forget we have limitations until we are brought up short by exceeding them.

A broken airplane seldom comes about because of one simple incident. It usually takes a combination of factors, such as a dark night, a gusting crosswind, and a tired pilot, acting in concert, to make it happen. Even a mechanical failure can usually be prevented from becoming catastrophic if the pilot has enough knowledge of the airplane's systems to isolate or work around the problem. Safe flying starts with *knowledge* so that the pilot will recognize what the equipment can and cannot do.

Acquisition of knowledge is an ongoing process, not merely a set of information to be learned with no further study needed. Aviation is a constantly changing business and, although the weather and the basic laws of flight are immutable, operational restraints and airplane handling procedures do change. The careful pilot will keep up with what's going on by reading the airport and online bulletin boards, consulting the *Aeronautical Information Manual* (AIM), catching the local NOTAMs from an ATIS broadcast, and yes, subscribing to the aviation press. The pilot who gets his or her ticket and stops studying only goes downhill—with serious safety results.

BEING HONEST WITH YOURSELF

Honesty plays a large part in developing a pilot's judgment. You must recognize that there are things you can't do—or at least shouldn't do—even if other pilots are getting away with it. Attempting a zero-zero takeoff in fog might be practical for the alert, rested pilot who has hundreds of instrument hours in that particular cockpit, yet foolhardy for the tired, rusty, or inexperienced aviator in a strange airplane. Be honest with yourself. There aren't going to be any traffic cops standing at the airport to hold you back from killing yourself. You're going to have to say *no* to yourself—by yourself. And, when pressures of your business or family are involved, you had best be a pilot first and everything else second.

Regulations, you see, are quite vague about what is and isn't safe operation. This is because pilots vary widely in capability, as does their equipment, and the regs are weighty enough without adding a chapter and verse for each type of operation. The FARs provide only the sketchiest framework of guidelines. They must be supplemented by our own restraints to provide adequate margins of safety. What's SOP (standard operating procedure) in New England just isn't smart in a Rocky Mountain meadow. IFR fuel reserves sufficient for the coastal regions had best be increased across the Great Plains. One might indulge in a bit more boldness when a glideslope receiver and marker beacon are installed, whereas a single-localizer setup demands higher-weather minimums.

So we all have limitations. I venture to say no two pilots will have exactly the same set of guidelines; your own personal limits will vary from day to day. We all have those days when we wake up tired and perhaps are fighting off the leftover flu bug or the effects of last night's party. It doesn't make sense to tackle the same tough situation under these conditions as we might on a clearheaded day.

One of the most dangerous circumstances is an airplane with two pilots and no captain. When a pair of aviators are of roughly equal experience, neither wants to throw in the towel and admit defeat. Often the person at the controls will continue on in the face of doubtful odds, secretly wishing he could quit, while the individual on his right silently wishes he *would* quit. Remember, airplanes are not run by committee decisions. Don't let anyone—another pilot, the boss, a controller, *anyone*—pressure you into exceeding what you know are your proper limitations.

ARRIVING AT LIMITATIONS

But how to arrive at a clear-cut decision about limitations—that's the question. Start by defining the mission. Is it a departure with no immediate need to return, or will it be necessary to come back later in the day? Will it be a strictly VFR trip, or do you feel capable of handling IFR conditions, in *this* airplane, in *this* area? Is there a hard-and-fast schedule to be met, necessitating alternate modes of transportation if the flight is scrubbed, or can there be more flexibility in departure and arrival times?

Having defined the mission, carefully consider the weather situation. If you don't understand the weather, don't fly in it. Ask the FSS specialist about the synopsis; match existing conditions with what was forecast, then consider whether or not you want to bet your life on what you've been told.

If you have seen this type of weather situation before, in this same area, you may have a fair handle on what to expect aloft. On the other hand, if you're a stranger in a strange land, watch out, and don't push your limitations.

Weather limitations depend on the pilot's experience, equipment, time of day, terrain, and alternatives available. By all means, always have an alternate everywhere along the line. Instrument pilots aren't the only ones who need alternates. Whether IFR or VFR, it always pays to know where you can go if conditions deteriorate or an emergency occurs. Never leave yourself without an "out." Knowing there is a nice warm airport a few minutes away helps avoid anxious moments and occasionally preserves an airplane.

Finally, after recognizing the task ahead and the weather to be faced, take a look at yourself and your equipment. You and the airplane are a team; neither can do the job alone. If you aren't feeling razor-sharp, keep out of tight weather and long, tiring flights. If the oil consumption is running a bit high, or the primary radio has been going to sleep occasionally, don't push yourself quite as far before yelling "Whoa!"

JUDGMENT

Most accident reports show the pilot exercised poor judgment at some point in the flight, perhaps before she left the ground, en route, or even at the last minute, just before the crunch. Judgment is not something you

Knowing whether to attempt a departure toward these storms requires not only a careful weather briefing but a willingness to use good judgment.

can buy, have subliminally implanted, or borrow from other people. It is something to be acquired from within yourself and it is absolutely vital for your survival. Kid yourself on the highway and you may get a ticket; lie to yourself in an airplane and you may get dead. Good judgment is simply recognizing that you do have limitations and resolving to stay within them. Poor judgment is allowing yourself to be swayed by the press of nonrelated forces to the point of compromising your limitations.

Good judgment is adding full fuel when it looks like the weather at the destination is falling apart. Poor judgment is pressing on because you need to be there and a fuel stop will take too long. Good judgment is driving a trip because the forecast wasn't proving out in the sequence reports; poor judgment is launching off "just to see how it looks" when you know that doing so will almost certainly require you to continue the flight. Good judgment is waiting an hour for another set of METARs to come out; poor judgment is taking off without checking weather, figuring on picking it up en route. Good judgment is asking a mechanic to look at something you suspect is a problem; poor judgment is putting it off until the airplane is in for its annual in a month or two.

You can't inject judgment into a vacant mind. It comes from within, tempered by experience and knowledge. Don't ignore the reality of limitations, and strive to possess a pilot's longevity insurance—good judgment.

SLOW DOWN TO SAVE TIME

A t any time during the chain of events leading to an accident, the pilot involved could have taken action to prevent the mishap. Obviously, he or she did not. We who were not direct participants in the calamity often ask ourselves, "Why would a guy do that?" If the pilot is still around to give us an answer, he might very well say, "I was in a hurry and I just didn't take time to...."

Haste in the cockpit is ingrained into us from our initial hour of training, when emphasis was placed on "staying ahead of the airplane" lest the machine seize the opportunity to run amok. However, **we must guard against taking satisfaction in the sheer rapidity of our movements around the cockpit.** Like that rat running a maze, a pilot can be trained to manipulate controls habitually, and if the airplane or the situation strays from the routine, a pilot's conditioned responses may not be the correct ones. A few examples will bear out my hypothesis.

HABITS AREN'T ENOUGH

A friend of mine who made a gear-up landing told me the accident was a matter of following habitual procedures. As he planned his departure before taking the runway, he intended to leave the gear down in the pattern. But after liftoff his hand went to the gear lever as it had for thousands of previous takeoffs. We can almost see his trained response bring up the gear, ease back the throttles, tweak the props into synch, and add a touch of nose-down trim, all by subconscious, learned action rather than by conscious thought. As he concentrated on the landing that followed, he knew the gear was down because that's how he had planned it before takeoff. The result was a perfectly routine procedure right up until the prop blades hit the pavement.

The pilot's problems began with haste, the need to make a quick up-and-around pattern, which diverted attention from the matters at hand, such as confirming gear-down. The airplane was being superbly flown by trained reflexes right to the point of impact; he never recalled raising the gear, because his brain was not involved in the action.

HASTE EQUALS WASTE

That haste can make waste was brought home to me rather forcefully by my first multiengine instructor several years ago. We were shooting touch-and-goes, and I was getting good at it. Every touchdown was followed by the same pattern; power up, flaps up, gear up, climbout, and blueline. We were flying an early Aero Commander with main gear that simply swiveled aft for retraction, and my automatic touch-and-go responses had gotten so quick that I'd unlocked the gear and moved the handle while we were still on the runway. As the old Commander began to settle the foot or so between its belly and the surface, my instructor hauled the yoke back in his lap to keep us airborne and delivered one of the few curses I ever heard him utter. "Don't get in such a big ******* hurry; take some time to fly the airplane" was the gist of his stinging lecture. My quick hands in the cockpit were not a virtue, they were a vice.

Even during moments of stress, take your time. Remember, if you have only one chance to get out of an emergency, take a second to make sure you're doing the right thing. All too many times a hot-rock multiengine pilot, faced with an engine-out emergency, has hastily pulled his levers, zip, zip, zip—and feathered the wrong engine. Yes, something needs to be done, but it has to be the *right* something, or you're not going to make it through the darkness to the other end of the tunnel.

PROPERLY ORDERED PRIORITIES

Whenever the situation demands positive, immediate action, you should establish priorities. You can't do everything at once, so take care of first things first. An instrument student making a missed approach will often bring in a touch of power, raise the nose a bit, and grab for the micro- phone while the aircraft is still settling, anxious to get his report out of the way. Forget the radio for a few seconds, until you have first set the power, stabilized the climbout attitude, and, perhaps, begun the turn toward the missed-approach fix, if an immediate change in heading is required. Then tell the tower you've missed, on the odd chance that they haven't already seen your pull-up.

Even the basics of airmanship require a certain amount of patience, a trait not always compatible with our desire to execute the privileges of pilot-in-command. A fast level-off from a climb, coupled with a quick power reduction and trim adjustment, will result in several hundred feet of altitude sinking away, all because the pilot didn't give the airplane time to accelerate. It may be an unfamiliar combination of high-density altitude and heavy load that causes the airplane to be a little slow in coming up onto the step, but whatever the cause, the pilot should have been watching and feeling the airplane, instead of trying to play up-tempo music on the cockpit controls.

STAY WITH THE AIRPLANE

There is a limit to an airplane's capacity to produce, and you should pace your actions accordingly. Even if you're trained to the point of lightning-quick moves, keep your brain engaged to make sure the airplane is keeping up with you. Don't pull the throttle and prop back to a reduced climb power just as soon as the airplane breaks ground; give the VSI time to build and grab a little altitude first. Even if you have noise-sensitive neighbors off the end of the runway, leaving the power alone for a few seconds and holding V_y speed will result in less perceptible noise impact than auguring across their rooftops at low altitude into a flat climb.

Trying too hard invites negative results. Back when I was beginning to learn the rudiments of glider flying, I would wrap the sailplane into a 60° bank at a minimum-sink speed, trying to stay in the hot core of a thermal. I usually wound up flying in and out of the stall burble, generating more sink through drag than the lift of the rising air could offset. An old-timer showed me how to do it, with a less aggressive technique: Maintain more airspeed or use a shallower bank to avoid the stall burble, and milk the overall thermal until you can gradually shift into the core. Even a little climb is better than a sink.

The biggest difference observed in flying with a truly experienced pilot—a wise old bird with years, not just hours, logged in the air—is the relaxed way he uses the controls. As you watch such a person fly, you see no wasted motion, no doubling back to correct lost altitude, no resetting of trim because of a hasty move. This pilot has learned to wait and watch the machine, giving it what it needs, exactly as much as it needs, just when it needs it, not before and not after. It doesn't matter that he has never flown this type of airplane over this route before; give him a few minutes to settle into place and he'll find out how to get what he wants from the airplane, with patience. If the situation is complex, he will probably ask you to answer the radio or look for an approach plate; he's setting priorities, and the airplane always comes first.

We can hear the distinction of experience each time we fly by analyzing radio technique. A newly rated eager beaver pilot blurts his call-up or readback at 150 words per minute with gusts to 200, a method that is sure to generate a "Say again" from the controller. The older, wiser head will talk at a rate compatible with the traffic situation, endeavoring to get his information across on the first transmission to save time in the long run. Strive to eliminate unneeded verbiage from your transmission, but keep your speed of delivery slow enough that a controller won't have to ask for a repeat or, worse yet, possibly misunderstand your intentions. Take a moment to listen after coming on a new frequency before transmitting; just because you were told to report doesn't mean it has to be done and completed within a 2-second void period.

By all means, stay awake in the cockpit, but make your actions studied and purposeful rather than just quick for quickness' sake. Fly in a state of relaxed alertness, establishing priorities to handle the busy times but never letting your hands run ahead of your brain. It takes less time in the long run and the airplane will be flown more professionally. You will wind up staying truly ahead of the airplane, thanks to the time saved by eliminating wasted motion.

STRESS MANAGEMENT

Everybody talks about stress and how it affects people; let's see how stress can influence a pilot. A little stress can be a healthy challenge, but a lot of stress can be tiresome. Excessive stress can be life threatening. Nearly all contemporary lifestyles contain stressful situations, and as modern-day pilots we are subject to considerable stress as a result of our endeavors.

Flying isn't always stressful. As we all know, boring holes in the serenity of the sky, far above one's earthly troubles, can be a soothing way to shed a burdensome load of cares. Unfortunately, we twenty-first-century types ruin a good thing by building additional stress into it. We often attempt to put our aerial toys to work as tools of transportation, which is a fine example of having our cake and eating it, too. But coping with the age-old perils of the sky while trying to adhere to a schedule is like turning up the heat under a pressure boiler.

Were it not for the effect of stress on their flying, we wouldn't concern ourselves with pilots who try to fly from point A to point B on a timetable. But when a person's ability to function effectively as a pilot breaks down as a result of his or her stress overload, it's time to look at the problem and see if there is a solution. For instance, if a pilot loses most of the preceding night's sleep worrying about whether or not she would make it to the big meeting tomorrow, she won't fly as well as she might were she well rested. As another example, the pressure to get home in time to punch the clock on Monday morning is just the sort of stress that has buried more than a few families when common sense gave way to unrestrained urgency.

SHEDDING STRESS

Let's try to understand what goes on in the pilot's mind when a flying problem arises; then we'll see if we can "unload" this highly stressful situation.

Psychologists say people handle threatening problems in one of two ways, both of which are quite effective. One way is to ignore the whole thing and block it out of the mind completely. It is particularly appropriate when there really isn't any individual action that can be taken to solve the problem. Averting a major war is a good example. You say, "It's out of my hands, so I just won't worry about it," and by dismissing it from your conscious thoughts it ceases to be a source of worry.

The second method of coping with a threat is to take all possible precautions in preparing for it, thus relieving yourself of uncertainty by saying, "There, I'm ready; let it come." Averting fears of a major war in this way might mean hoarding a year's supply of food or selling stocks in vulnerable industries. Both types of reaction can be carried to ridiculous extremes, but they are typical of how we handle daily life.

In flying, you can't ignore all problems, but you can ignore the ones you can't do anything about. If you are 500 miles away from home and really want to start back, but one member of your party isn't going to be ready to leave until tomorrow, you may as well stop worrying. You're not going anywhere. Staring out the window won't do anything to hurry up the departure, so put the flight out of your mind and enjoy whatever pleasures the local attractions have to offer.

From the beginning of flight training, you are instructed to leave nothing to chance. Flying is a terribly unforgiving occupation to those who make dumb mistakes. Therefore, taking appropriate precautions would seem to be the more useful reaction in a highly stressful situation. This increases your workload and spoils the fun of a spontaneous flight, but it must be done if you're to avoid dangerous situations that could create even more stress.

PREFLIGHT STRESS RELIEF

Let's divide our actions into two areas: actions related to preflight preparations and actions undertaken in flight. As you approach a trip that has been scheduled for some time, stress mounts up naturally. Will the weather be up to your minimums? Will the airplane you have reserved get back on time? Is there enough daylight to get there before dark? Can you afford to miss another day of work if you don't get back? And so forth.

This set of circumstances brings us to the old go/no-go decision. How you handle this stress maker depends on how closely you have to adhere to the flight plan. Perhaps the time of departure isn't so critical, if the airplane isn't rented for a specific block of hours. If the weather is bad in the morning, leave in the afternoon. If you can't get organized in time to make it today, fly tomorrow. Tell everybody involved that's the way you're going to play the game and you can stop worrying about the timetable.

Even the destination need not be cast in concrete. If you've promised the kids a ride to West Point and you're fretting about the forecast in that direction, plan an equally entertaining trip in another direction for the little dears, and cease that pacing back and forth to the window. By going one step further with a third, ground-bound plan that you can press into service if all flying must be scrubbed, you no longer have anything to get uptight about, having covered every eventuality.

But what about those *really* important trips, the ones you've just *got* to make? They're sure to cause ulcers unless you assess the situation realistically. No means of transportation is flukeproof, and that goes especially for general aviation. Even the airlines occasionally must cancel flights because of weather or mechanical malfunctions. If the trip is important, simply set up another means of getting there if you're not able to fly as planned—fill the car up with gas, make an airline reservation, hire an instrument-rated pilot, or arrange a rain date. When you've done everything you can, go to sleep and forget about it.

If you can make some preparation to deal with interfering problems, do so. If you can't do anything more, forget it! Worry never parted the clouds or calmed a crosswind. For the must-be-there trips, set up a final decision time—the last possible moment when you could switch to the car or run to catch the airliner. If your path is not clear as that hour approaches, switch to your alternate and don't look back. I've driven a lot of miles while cursing

Stress on the pilot can mean stress on the airplane. This Navion's pilot forgot his wheels until he was nearly down; a last-minute go-around flattened the prop tips.

the sunshine overhead, but my decision time had come when the clouds were still thick. If you don't have the minimums you need, *don't* fly. Never believe a forecast until it starts to happen.

IN-FLIGHT STRESS RELIEF

Okay, **you've taken every possible precaution before getting into the air.** Now you're airborne and a different sort of stress sets in. Flying is a time-sensitive game; you have to be *somewhere* before the fuel runs out and you must keep tabs on your progress to see if you'll make it. Before takeoff, the passage of time was only an inconvenience, not a threat. In the air, time is of paramount importance.

Determine the source of any in-flight stress you feel building. If you're concerned about the sound of the engine, note the readings of the temperature and pressure gauges and check them periodically. No change usually means no problem. Troubleshoot by switching mags, trying the carb heat, and adjusting the mixture. Then quit worrying. If the situation deteriorates, just head for that alternate airport you've always had in mind. Be confident that you can handle the problem in an organized way, and stress can be managed.

Is the weather en route different from that forecast? That's a stress maker as old as aviation. Get hold of Flight Watch on 122.0 and find out where the weather is still good; there's your out. Then check to see if the observations ahead are up to your standards. If they aren't, you've just made a decision to divert, *regardless* of how important the trip may be. Remember, "must-fly" trips lose that status the moment you're airborne. You should always divert if safety demands it. Because you've done all you can, ignore the stress of missed meetings or canceled appointments. If the weather falls apart, don't risk your life to preserve a business schedule.

Never let stress build to the point of interfering with the rational thinking necessary for flying; doing so just *invites* an accident because the airplane lacks a coherent pilot. One corporation I am familiar with has a simple rule for its businesspeople-pilots: If they are not comfortable with their flight situation, they are to put the airplane on the ground and work out their schedule in the airport lounge. That attitude saves a lot of lives by allowing the pilot to shed stress like a dog shaking off water. We should all adopt that policy.

CONSIDERATE OPERATIONS

There was a time when pilots could fly pretty much as they pleased. As long as they stayed within the regulations, any public nuisance resulting from their actions could be smoothed over by the offer of a free ride. After all, pilots were supermen, to be given a trifle more slack than ground-bound mortals.

Alas, those halcyon days are gone. Your pilot's license now scores few points at a cocktail party ("Oh, really? Then you must know my brother-in-law; he flies, too"), and it has almost no armoring effect against the slings and arrows of outraged citizenry. As Rodney Dangerfield said, we don't get no respect. We're going to have to consider the effect of our actions, as perceived by others, if we want to count on the public's support in the clinches.

DEPARTING GRACIOUSLY

Consideration of others begins with the start-up. Strive for a low-rpm light-off and warm-up; 800 rpm will keep an alternator on the line and hold a little suction pressure for the gyros, and it will move a lot less dirt than 1,200 rpm and keep the ramp quieter as well. When taxiing, use minimum power and brake while moving to the runup area, and leave the strobes off so onlookers won't be blinded. A 1,500-rpm taxi adds to engine wear, shortens brake life, and is generally regarded as a sign of a poor pedigree. Then too, passengers usually don't care for squealing, swerving, high-speed taxi runs.

Limit your use of power to the few seconds necessary for mags, suction, and prop tests; it's amazing how far the sound of a runup will carry on a still summer evening. If you have first-time passengers on board, keep them informed about the sequence of events without alluding to impending doom; explain that the pretakeoff checks are done to make sure everything is in perfect working order. Do not say things like "If this mag quits the other one will bring us home" or "Looks like that'll be enough fuel." Exude confidence. Oddly enough, most new passengers don't like to see the captain perusing a checklist ("Can't he even remember what he's supposed to do?"), so use it but keep it unobtrusive.

Brief your passengers on what will be happening before you kick in the afterburner. Tell them it will be somewhat noisier than a car, that a turn will follow the liftoff, and that any low-level turbulence will be topped shortly. Take a good, vigilant look for traffic, even if you have a controller's blessing. If operating uncontrolled, don't pull out in front of an airplane on short final or close-in base leg unless you're in contact with the pilot; the FARs give the landing aircraft priority for very good reasons. You may precipitate a go-around even though you feel there's plenty of room.

When using the radio, think about the other fellow. Check that you have the volume turned up before you talk, and don't start blasting away right after dialing up the channel; listen for a second or two, in case there's a comeback in progress. If you have manual squelch, open it up a little on the first contact to make sure it isn't filtering out a distant reply, then adjust it after you raise the tower. Think before you talk, don't compose on the air, and avoid nonproductive conversation.

KEEPING THE PEACE

More public ire is raised over departure procedures than any other flight operation. If there is a preferred takeoff path, use it. If there isn't, create one; flying low over housing developments usually can be avoided. If an extended straight-out will get you around the edge of town, maintain the runway heading. Comply with the tower's wishes if the controllers express any preferences, but go further than that; in the interest of future harmony, watch for noise-sensitive areas on your own. For instance, I frequent one uncontrolled airport that gets a fair amount of jet traffic and 90 percent of the big-iron drivers will make a right turn out, smack-dab over the town square, because that's the logical way to head for the VOR. The management should pass the word, but doesn't because it's afraid of hurting Chairman Big Dome's feelings. If John Q. Public gets riled enough, Big Dome may find that airport gone someday, so a word to the wise would be in order.

Departure technique should place as much altitude under the airplane as possible, as quickly as possible, to minimize the exposure at high power settings. Some pilots mistakenly use partial power for takeoff, or reduce power abruptly after liftoff, attempting to lessen noise. The result may be exactly the opposite; reduced power means less altitude will be gained by the runway end, and as the edge of town is reached the neighbors get a big dose of pistons swapping holes. As long as your passengers can tolerate the

The pilot of this Piper Comanche knows to avoid low flying around the power plant beside the lake, out of consideration for security conscious citizens.

high deck angle, rotate on schedule to achieve V_y at 50 feet and leave the power at full bore for a minute or so, hanging the airplane on the prop to leave the airport vertically rather than horizontally. No certified engine will be harmed by this short full-power run; as a matter of fact, engines sometimes wait for the first power reduction before coming unglued, so there is every reason to leave the throttle alone for a while.

I would make one exception to my wide-open rule, for airplanes powered by the 300-hp Continental IO-520, rated at 2,850 rpm for 5 minutes; that big howler rattles the dishes halfway to Cucamonga as the prop tips go supersonic. If I owned one, I would have a mark inscribed on the prop vernier shaft so I could limit takeoff rpm to 2,700 unless the strip were supershort. Only 15 hp is lost by derating the engine in this manner, not a significant amount in normal operation. One hopes that keeping the prop speed down will hold the neighbor's temper down as well.

The maximum-rate climbout should be held for the first 1,000 feet of altitude, at which time a gradual transition is made to a more customary climb angle. If this can be done within the airport boundary, the initial power reduction will be made before overflying any residential areas. Should there be no alternative to departing over those poorly placed houses surrounding small airports, make the reduction from takeoff to climb power as soon as you think it safe—perhaps at the end of the runway.

Airplanes with fixed-pitch propellers seldom need power reductions for climb, because they customarily are equipped with cruising propellers that limit rpm to well below redline at low airspeed. Airplanes with constant-speed propellers are typically given a power setting of about 80 percent for climbout; prop speed is the critical factor in controlling noise, and at elevations above 4,000 feet MSL, the throttle need not be pulled back before the prop control, because manifold pressure will not be greater than 24 inches in most normally aspirated engines.

Common courtesy dictates that you inform the control tower of your desired direction of flight when you report ready, so the staff can plan accordingly. At uncontrolled airports, taking a shortcut onto course by angling around into the touch-and-go pattern is poor manners; if you can't gain enough altitude to clear the pattern, you should swing out a mile or so.

EN ROUTE NICETIES

Consideration of others continues in the en route phase of the flight; don't assume your passengers are capable of the same bladder-busting enduros you can make. For the nonveterans, a 2-hour hop is plenty long enough, and even that should be cut short in turbulent conditions. Try for smooth air for your passenger's sake, both in timing of the flight to avoid the roughest part of the day and in choice of cruising altitude.

Don't be guilty of drifting upward a few hundred feet at a time with the rising turbulence level; breaking the odd/even cruising altitude rule can easily result in a head-on conflict at a closing speed of 300 mph, so if you're going up, make it a full 2,000-foot change. If your airplane ascends in an extremely nose-high attitude, break up the climb with shallow turns, both to clear the airspace ahead for oncoming traffic and to create movement in the other pilot's field of view.

As you progress along your route, note significant unreported weather and make a pilot report to Flight Watch. Your report may save someone else from wasting fuel when the pass is closing rapidly, and knowing a clearing line's position can ease a worried pilot's mind. If we keep such information to ourselves, everyone will have to discover the truth separately.

Remember to put as much thought into the arrival as you did the departure. By planning the letdown, you can leave the power at a low cruise setting when you level off for the traffic pattern, instead of dragging along at cruise power after letting down early. To eliminate the snarl of redlining rpm, don't run a constant-speed prop into low pitch until the power is reduced for approach.

ARRIVAL COURTESY

Get your communications act together early in the game, listening to ATIS or the tower's chatter to get the numbers, then checking in at a well-defined reporting point. Keep all conversations short, whether with tower, FSS, or Unicom, but talk slowly enough to preclude misunderstandings. Double-check frequencies, most particularly at uncontrolled fields; look for the CTAF (common traffic advisory frequency) symbol on the chart to make sure you're on the correct Unicom or Multicom.

Bear in mind the needs of others during the approach phase of the flight; if you see traffic holding for departure, fly a short approach rather than take a grand tour of the countryside, so the pilot on the ground can get her overheating engine into the air. Try for a turnoff at the first intersection to eliminate the need for a 180 on the runway. If you have traffic following, keep your speed up as long as possible and clear the runway promptly; creating a go-around situation needlessly is poor manners. Never, *never* make a right-hand pattern just for your own convenience at an uncontrolled field with standard patterns.

Use some discretion when parking your bird; the lineman may be busy when you wander onto the ramp, if such a person is employed there, casting you on your own devices. Pick a spot that lines up with the other airplanes and doesn't block hangars, taxiways, or the fuel pits. Before you run off to town, hunt up an official person and ask if it's okay to park there for the appropriate length of time. You may be occupying someone's private slot with passion-pink wheel chocks and custom-fitted tie-down ropes.

Your consideration of others encourages the kind of discipline needed to make you a superior pilot. The aviator who charges ahead without regard to the effect of her actions probably will bring herself to grief someday, because of the lack of restraint in her personal flying habits. Be a decent sort; it'll improve everyone's frame of mind and make you a better pilot.

SENSIBILITY

In aviation, as in all fields of endeavor, some things are just not discussed in school. Training texts will contain pat, cut-and-dried answers to specific questions, leaving students to learn later in their careers that some situations just don't fit the textbook examples. Checkrides are passed by reciting the standard catechism drilled into the novitiate's head by curriculumbound instruction. The FAA handbooks and manufacturers' approved manuals simply ignore nonstandard, deviate behavior, assuming that no one will ever stray from the straight and narrow. Alas, there will come a day when an emboldened young aviator will wonder "What if..." and, because he was never told specifically why he shouldn't experiment, he might try some forbidden act just to see what happens. And *that* my friends, is an open invitation to disaster.

BUT WHY NOT?

We live in a questioning, testing age. Today's educators are encouraging students to ask "Why" or "Why not," and if no one can give them what they consider to be a straight answer, they are encouraged to attempt to find out for themselves. This may be healthy, but it can lead to a lot of bruises in the course of experimentation. The point is, we can expect the current crop of young pilots to try some stupid things, just like yesterday's pilots. We can only hope that they won't ignore common sense.

Let me be the first to disclaim any intentions of encouraging illegal or unsafe acts. Hopefully, all pilots will refrain from stepping outside the guidelines of safe operations as set forth in the regulations and approved flight manuals. However, there is a real world where aviation exists as a self-regulated activity for the most part, all the carefully worded rules notwithstanding. To protect the interests of their employers, government bureaucrats and company lawyers will write only broad, generalized statements containing simple restrictions with no incriminating illumination.

When an airplane is placarded "intentional spins prohibited," its pilot has no way of knowing if the aircraft was found to be unstable at some

If a pilot decides to tackle this much crosswind, he'll learn the reason *why* behind the POH's maximum demonstrated crosswind component.

point during the spin, if it would recover safely but not quickly enough to satisfy certification requirements, or if the manufacturer simply didn't want to assume the liability of endorsing spins. So if a pilot is feeling his or her oats someday, and decides to try a spin, the outcome may be very much in doubt. But human nature being what it is, a blanket prohibition is just not enough to restrain some bold but unknowledgeable individuals.

It is foolish to expect the situation regarding specific information to improve; society is advancing ever deeper into a "sue thy neighbor" philosophy, and open honesty has now become so unexpected as to be suspect in itself. Therefore, pilots bent on survival must look beyond the letter of the law as written and apply the timeless guidance of common sense. We are going to examine some less-than-legal operations that are often undertaken, officially sanctioned or not, and will try to offer sensible restraints to the pilots bent on self-destruction.

BOONDOCKS OPERATION

Consider the off-airport landing—legal, but it just isn't done, right? Well, a little observation reveals that it *is* being done—by bush pilots, glider pilots, crop dusters, and a few thrill seekers. After all, it doesn't take too many touchdowns on some poorly maintained dirt strips that are officially recognized as airports to convince a pilot that a cow pasture can't be much worse. So the next time our pilot decides to visit Uncle Ben's farm, he or she may try landing on the back forty. Having gotten away with it once, the pilot is likely to try it again and again, until finally coming to grief.

Rather than expressing disapproval of all landings on unpaved surfaces, as is the usual practice during training, let's apply common sense. Off-airport landings are (1) terribly rough on airplanes, particularly the more modern designs; (2) risky, much like driving at night with your headlights off; and (3) probably outside the coverage of your insurance policy. Therefore, when considering operating from anything other than a regularly used airport, you should first *walk* the entire surface, not just drive over it or fly over it, and certainly you should not take anyone's word for its condition. Count your paces to verify the length and look at every square foot of the area to be used; it takes only *one* unseen rock or gopher hole to ruin your day. If you decide to land there, at least you'll know what you're getting into. Use common sense and you'll probably find yourself sticking to normal airports.

ILLEGAL AEROBATICS

Now on to another subject seldom discussed, merely forbidden: acrobatic flight in normal-category airplanes. Ground school prohibitions quickly forgotten, a new pilot soon discovers that (1) ordinary utility and normal category airplanes are perfectly capable of being flown into some unusual attitudes, especially if entry speeds well above maneuvering speed are used, and (2) it sure is a lot of fun to see the horizon from all those different angles. Unfortunately, every so often someone takes an airplane apart trying to prove he or she is as good as Bob Hoover, little knowing that Bob Hoover wouldn't dream of attempting his routine without a lot of careful experimentation first.

Again, use some common sense. If you're going to throw an airplane around, you need lots of altitude, a very strong machine, and some training in recoveries from unusual attitudes. It makes no difference that the self-taught aerobat may have gotten away with a flip-flop or two in a stock machine; eventually he or she will lose control and fall out of a maneuver, possibly exceeding the airplane's stress limits. At the least, a bad scare will be experienced; at the worst, the pilot will wake up dead. Common sense should restrict you from attempting aerobatics in stock flying machines, even if you choose to ignore the regulations.

THE BUZZ JOB

Next, I'd like to see the hands of all those who have *never* buzzed a friend's house to let him or her know you were in the area. I thought so.

It clearly states in the regulations not to fly lower than 500 feet AGL, except over open water or vacant desert, yet most of us have been tempted into sneaking a fast pass down over the girlfriend's house, possibly getting away with it often enough to make it a habit. Buzzing is both risky and illegal, but if you're going to do it, use some common sense.

Don't loiter around sightseeing at low altitude. Make a cruise-speed dive, pull out over the objective, and get back up to a reasonable altitude immediately, in case the engine should pick that moment to pack up. Don't go down behind the trees or outbuildings—that isn't going to impress anyone but the grim reaper. Stay well above any obstructions. Don't let the airspeed drop below normal climb speed, and avoid crop duster turns down on the deck. One pass should be sufficient for a day's thrill, if you're determined to show off your boneheadedness. Remember, a lot of nonflying people are negatively impressed by a buzz job, and they've got the votes to ground us if their ire is sufficiently raised.

FLYING TANKED

Next, **let's talk about booze and flying.** Never going to fly while under the influence, are we? The only problem is deciding when we're under the influence. A person wrapped in a warm alcoholic glow is hardly the one to determine his or her own fitness to fly, yet that person is the one doing the deciding in nearly every case. The regulation specifying 8 hours from bottle to throttle is only marginally better than no guideline at all, because it makes no allowance for the rate and amount of consumption or for body weight, all of which affect sobriety.

The fact is, there's no such thing as too little alcohol to have an effect. Whenever alcohol is consumed, there is always a depressing effect, which is enhanced by altitude. Even *one* beer or cocktail robs you of a portion of your faculties, unbeknownst to you, so only your common sense of self-restraint will prevent you from flying while under the influence. Don't trust your judgment after you've already imbibed. If flying is a major part of your life, you may want to seriously consider becoming a teetotaler, substituting the joys of flying for drinking, if you will. By staying on the wagon, you'll improve your chances for a long flying career, avoiding possible heart, liver, and brain damage. If you insist on maintaining what has come to be accepted as an active social use of alcohol, avoid any flying on the "morning after." Stay on the ground for 12 hours after moderate drinking, to be more safe, and for 24 hours following a real elbow bender. Remember, common sense isn't likely to be working when you're slightly juiced, so make your decisions now.

A LITTLE SOMETHING EXTRA

Overloading is yet another dark area normally left untouched by training. Even the most inexperienced students can't quite believe that an airplane will refuse to lift off if it is merely 1 pound over gross, but they are seldom told how much might be gotten away with in terms of overloading. When entering the real world they are forced to decide for themselves when to offload fuel or baggage to maintain safety—at 1 pound over gross, 10, 20, when? The reasons for avoiding overload are well-known from ground school—the landing gear structure may be overstressed, operating limitations are more easily exceeded, and the performance charts for the aircraft become so much worthless paper. However, from a practical standpoint, few pilots are going to get excited about a 5-pound overload.

So how much is *too* much? New pilots often ask, "A 10 percent overload beyond gross weight is all right, isn't it?" *No, it isn't.* Even 5 percent will noticeably degrade aircraft performance; on a 2,500-pound airplane, that would be an extra 125 pounds. Unless the situation is an emergency, requiring a compromise to avoid greater risks from another source, 1 or 2 percent of gross weight is quite enough overload to ask the airplane to carry.

There are any number of other gray areas where the pilot's judgment must fill in the gaps between regulations. As previously stated, aviation is largely self-regulated, and if you want to do something stupid, few constraints other than your common sense exist. Please, if you feel an overpowering urge to throw caution to the winds, research the matter first to find out why the regulations or limitations prohibit such an act. Then listen to your common sense, rather than the siren song of temptation.

CURRENCY

Every so often someone asks if I'm current in a Comanche, or a 310, or a SeaBee, or just about anything. Although I seldom give a straight yes-or-no answer, tending to indulge in replies of "More or less," I have come to wonder just what they would really expect a normal airplane-hungry pilot to say. If you were being given a chance to fly an airplane of your heart's desire, and you thought you had a reasonable chance of getting it up and down without bending anything, wouldn't you probably say "Sure, I'm current"? I mean, who's going to say they're not current when there's flying to be done?

A case in point began with a phone call I received several years back. A chance acquaintance had left his Aztec weathered in at my field, and after continuing to his destination and tending to business, he needed his bird in a hurry. Could I bring it down to him, if I was current in an Aztec? Heck yes, I'm current, I mean, sure, let's go. You get the picture. What I said was, "I'll be there in an hour, if I can get it started." My previous Aztec experience was a demo flight 2 years earlier, a total of half an hour, but I remembered the airplane as being docile and straightforward, and I saw no problem.

Bear in mind that I was doing nothing illegal; I was multirated, there were no passengers, it was strictly a gratis, VFR ferry flight. His banker and/or insurance company would have no doubt been uneasy, but I felt confident that I could make the short hop because I had done considerable homework on Aztecs and there was no weather or darkness problem. All went well, he was happy to get his airplane, and I was glad to get the time.

WHEN ARE YOU CURRENT?

Now, back to the original question: When are you current? That depends on how demanding the flying is to be. After thousands of hours in a Cessna 150/152, I feel reasonably sure of myself in one of those airplanes, and I could probably do about anything I wanted to with full peace of mind. But an airplane that I have only had an opportunity to fly a couple of hours is still not part of me, and I would consequently restrict my operations until I had more experience. So currency is up to the individual, depending on circumstances.

The FAA, wisely recognizing that both pilots and airplanes differ widely in their qualifications and abilities, stays pretty far away from this question of currency. To be legally current, the pilot must be rated in the category and class of aircraft, have his medical and flight review in order, and have his high-performance airplane endorsements if he's flying an over-200-horsepower or retractable-gear airplane. If he wants to carry passengers, he needs three landings in the same category and class within the preceding 90 days. That's it; the rest is up to common sense.

The rules have to be flexible, because they must be written to cover all pilots. If we sought a regulation that would prevent a low-time pilot from jumping into an unfamiliar airplane in the middle of the night and flying into hard-IFR weather, it would also unduly restrict an experienced pilot who hasn't flown recently from using a simple airplane for a local pleasure hop. Further, some 250-hour pilots are worthy of the utmost trust, whereas some 3,000-hour aviators are dangerously overconfident and underskilled. There is a limit to the number of laws one can write and still expect to have them obeyed. In most cases, the owner's insurance company will take over where the FAA leaves off, and the keeper of the airplane's keys ought to go even further when setting forth restrictions on the airplane's use.

Therefore, in answering the question, "Are you current?" you had best counter with "Current for what?" A test hop around the local area, or a 1,000-mile cross-country trip? Nobody can make you suddenly, magically, current for every operation the airplane is capable of; only the passage of time while strapped to the seat cushion of the airplane can do that. You can lie your way into an airplane if you want, but it's your behind that's on the line when you can't handle the situation, so it pays not to overstate your qualifications.

START ON THE GROUND

Okay, **let's become current enough to fly a strange airplane** in VFR daylight conditions, then proceed into IFR/night flying later. Everybody wants to charge right out and get into the air, figuring that's the way to get current in a new bird. Not so fast. Those days are over; flying the airplane is now the simple part. Before you ever start the engine you need to spend an evening poring over the airplane's handbook and papers, making sure you have a working knowledge of the systems and recommended procedures. Then sit in the cockpit for a good long while, learning the position of controls and indicators with the handbook in your lap.

Find a good check pilot, CFI or not, who knows the airplane well; pick her brain for all it's worth. A CFI will have the advantage of being used to implanting procedures into empty minds, but she has to know what she's talking about before she can do that. With your preflight study, you should be a fertile field for such oral instruction.

The actual period of flying time necessary to make you competent to fly an airplane as pilot-in-command varies all over the place. If you have never flown anything but a Cherokee 140 in your entire 75-hour career, and that was a year ago, trying to check out in a Cessna 182 is going to take a while. On the other hand, if you have 1,000 hours of bona fide time and are regularly switching around in Cessnas from 172s to 210s, climbing into a strange 182 is no big deal, because you already know how Cessna does things. One gets used to changing mounts after a while, quickly finding the benchmarks of any new airplane to make an easy transition, but it's an acquired skill. (A note to such seasoned and confident switchers: Don't get too complacent, for they do change things on you, like using knots on the airspeed indicator or putting in a 100-octane engine instead of an 80-octane one.)

A checklist is vital to the transitioning aviator; without one, it is all too easy to omit some vital preflight check in the heat of battle, such as setting the directional gyro. However, after running the handbook's checklist, make a double-check of common items from your own memory (fuel, con-

Being current in recent experience in order to fly this Partenavia P68 involves more than just three landings in the last 90 days.

trols, instruments, trim, etc.) to make sure an unfamiliar checklist didn't lead you astray.

TAKE IT EASY

Naturally, for the trial flight you should select a day without wind or weather problems. Keeping the loading light and well within center of gravity (CG) limits is also important. A liftoff speed is selected on the basis of the stall speeds quoted in the handbook, and a rotation will be made just before reaching that figure, without actually trying to pull the airplane off. Let it run until it's ready to fly, then accelerate to best rate-of-climb speed after breaking ground. Just hold a heading and fly for the first few seconds—don't be in a big hurry to cut the power back or retract the gear and flaps. Keep up with the airplane and make all adjustments with a little forethought.

A 500-fpm cruise climb is usually desirable after the initial departure, going up to 3,000 feet AGL or more before leveling off to experiment with the airplane. Keep a watch on temperature gauges and other vital signs to become familiar with the power plant's performance. When leveling off into cruise, leave the power up for a few minutes to build up speed quickly, then make final trim and power adjustments for cruise. Note the sight picture over the nose in a level-flight attitude, then roll into some good medium-banked 360° turns to see if you can get the feel of keeping altitude by eyeball.

Slow flight is next, starting with a clean airplane. Work down to the point of the stall warner's indication, then, using little or no power, try a straight-ahead stall to ascertain the minimum flying speed and the aircraft's stall characteristics. Don't prolong the stall; recover promptly, at least until you are sure the airplane has no bad habits. Lower the flaps and/or landing gear and perform the same slow-flight and approach-to-stall drill. Add 30 percent to the stall speeds observed in clean and dirty configurations, and make these 1.3 V_s figures your minimum flying speeds until you have gained familiarity with the airplane. Bear in mind that you'll need to compensate for position error in indicated airspeeds, as shown in the handbook's performance section.

Take time to cover all aspects of the airplane's systems while in an actual flight environment. Try leaning procedures to see how much the mixture knob comes back; find out where the emergency gear-down handle is located; see if the radios are operable and accurate. If an autopilot is installed, try all modes to see if you understand its operation. Look around to see where the airplane's blind spots are located so you can be especially watchful for traffic approaching from that direction.

FIRST LANDING

Find a comfortable power setting for a 500-fpm descent at cruise speed and trim, and make the proper airspeed reductions before dropping the gear or flaps. One-and-a-half times stall speed should be a comfortable traffic pattern speed; plan on having a good long final approach to give yourself time to get squared away before touching down. Stabilize the approach at 1.3 V_{so}, then reduce power cautiously as the ground comes up to avoid falling through a flare with a heavy, unfamiliar airplane. Wait out any ballooning tendencies and ease the stick back slowly in small increments; never get into a pushing and tugging match with a strange airplane on landing. If things get out of hand, give it full power and start all over again.

Brake cautiously, once down, in case the airplane is resting only lightly on its feet and a wheel tends to lock up. Keep the stick full back during the rollout, and raise flaps when all their effectiveness is diminished, making sure you move the right control. Shoot more landings until you are either satisfied or disgusted, but do make them full-stop if at all possible; touch-and-goes do not give the full range of changing control pressures and can put the new pilot behind the airplane if she's not alert. Try various flap settings and power combinations during landings, seeing what works best for this particular airplane. Some aircraft need power right down to the ground; others float eternally if power-off dissipation of speed isn't complete before entering ground effect.

Having proved your prowess, taxi back to the tie-down and shut down, securing the aircraft against wind or prop blast. Now are you current? Hardly; you've just proved you can keep the airplane from biting you under ideal conditions. It is up to you to get out and fly the bird until you are thoroughly familiar, then try it out under less-than-perfect wind or loading situations. After much more experience, go on to IFR and night operations when you are utterly sure of your equipment and your mastery over that equipment.

When are you current? When you have ceased to be surprised by something the airplane does or doesn't do, not before. Let's face it, we can never be truly current enough in a flock of airplanes, but we should resolve to make every flight a step toward ultimate mastery of each. Don't lie to yourself about being current; you can fool everybody but the airplane.

YOUR OWN
AIRPLANE

FLYING CLUBS

A well-organized flying club would seem to be a very worthwhile concept. Pilots in the United States outnumber airplanes by 3.5 to 1, and the ownership of lightly utilized airplanes has grown even more financially unsound as the costs of upkeep escalate. Why, then, are there so few ongoing flying clubs in evidence? The changing moods of society might have something to do with it, and the time might be right to revise the prevailing attitude about clubs.

The traditional flying club is comprised of pilots who fly largely for fun, none of whom could afford to own the airplane(s) to which the club gives them access. Sharing the fixed costs in common, the club members pitch in for ownership chores with a spirit of camaraderie, meeting occasionally to elect officers, transact business, and have a good time.

Modern-day life seems designed to discourage group activities of any sort. We live in compartmentalized rat races, rushing to our next appointment past the person with a car's flat tire, struggling to make enough money to pay the bills. Who needs another meeting? Neighbors are people who moved in last year that we've never met, whose kids are probably on dope or something. Why should a club concept occur to us at all, when fewer and fewer families even get together to eat a meal?

With encouragement from social planners and a few suit-seeking lawyers, we've also become vastly less tolerant of anyone stepping on us, which makes us reluctant to share something as personal as an airplane. Cooperation can be taken as a sign of weakness, and the weak get eaten, so we tend to avoid situations that require getting along with one another.

LOWER AMORTIZATION

Maybe there is a time for all things, and **this could be the time to revive the once-popular flying club.** Individual ownership leads to poor utilization, and poor utilization is the biggest factor in the forced sale of an airplane when the bills pile up. If pilots band together to put 500 hours per year on an airplane, however, the per-hour amortization drops to reasonable levels.

Certainly, you can always rent. Or can you? Rental airplanes are not as plentiful as pilots would like, especially when FBOs cut back on their fleets to trim overhead. Rentals are primarily trainer-type aircraft; even if quite a few pilots would like a high-performance airplane to rent, few FBOs in the area will want to put one on the line. A good club can fill the yawning gap between rental and full ownership, if only we learn to accommodate each other's desires.

A club requires somewhat more from the individuals involved than just stepping up to a counter and saying "gimme the keys." Few clubs have the resources to order a wash-and-wax job done, or a $100 oil change. Taking care of the chores means direct, personal effort, not a bad ingredient for encouraging a spirit of togetherness. An evening's work party, followed by a little hangar bash, is good for the soul. On the other hand, if you prefer to "fly 'em and forget 'em," the club route isn't for you, and you must pay for the privilege.

It's important to remember what a club will do, and what it will not do. It will provide members with improved access to a familiar cockpit. An FBO permits you to fly whichever airplane happens to be on hand, if and when it is available, in whatever condition it was left by the last customer. A club member flies his or her *own* airplane, and must contend with only a limited number of other persons when he or she wants the keys. If the club accepts only nonsmokers, the next user knows there won't be any stale cigarette smoke in the upholstery. And the N-number on the panel will be the same every time.

What the club won't do is save buckets of money over rental fees, not unless the club members are willing to perform much of their own work, perhaps with an A&P as one of the members. For the occasional pilot, full FBO rentals still provide the cheapest way to go flying, because there are no monthly fees or special assessments. What the club does do is lower the cost of ownership; the per-hour cost of a personally owned airplane with a 50-hour annual utilization rate is many times typical rental rates, but putting five 50-hour-per-year owners into a club with one airplane breaks the bills down into manageable portions.

PITFALLS

Unfortunately, there are pitfalls aplenty, which scuttle many flying clubs. The aforementioned personality conflicts break up the best-intentioned associations, so club members have to be willing to give up rather than get even. Understanding the other fellow's problems is necessary, particularly when you can't leave on your vacation because Charlie wasn't able to return on time. The best defense against infighting is a good impartial set of bylaws; the bylaws run the club, not a petty clique or an individual

Flying club members must participate in their share of cleanup and maintenance chores, including verifying fuel status, in order to keep the airplanes in flight-ready condition.

tyrant. And bylaws must be enforced in order to avoid a precedent of letting Ol' Joe get by once or twice before cracking down.

Financial troubles beset all too many flying clubs, usually because the members deluded themselves about the true cost of keeping an airplane. It's wise to take in a former airplane owner who knows what to expect, or at least seek such a person's counsel. Once the facts are clear, the dues and fees structure must be set to cover the costs plus a contingency fund.

There are several types of club arrangements, some with a vested interest giving the right to sell one's share of the airplane(s), some with only a token membership. The first type might be formed by four friends, who put up a few thousand dollars apiece, finance an airplane together, and split the payments. The latter would be a more formalized, perhaps incorporated, group with a high turnover rate among a larger membership, none of whom actually own a plane but who pay monthly dues to run the club.

MONTHLY DUES STRUCTURE

To keep things on a sound financial footing, **all clubs need a monthly dues structure that will cover fixed costs,** so they won't be dependent on an hourly flying fee. Taxes, mortgage, insurance, storage, and periodic maintenance reserve must be paid whether the bird flies or not, so everybody gets assessed his or her part of the bill. The per-hour fee, on the other hand, takes care of items that are direct operating costs: fuel and oil, tires and brakes, oil changes and 100-hour inspection expenses, plus an additional donation to the engine overhaul reserve.

A worthwhile ploy is to bill members for 1 hour of flying time with the membership dues; they lose their prepaid hour if they don't fly it in the forthcoming month. This may ensure that at least a minimal level of proficiency will be maintained. Poor skill levels of inactive members has brought many a club to its knees when its assets were wiped out by an inept but card-carrying member. The bylaws should make provisions for a checkride with an elected safety officer or CFI if a member doesn't fly for a long period.

FBO-sponsored clubs can be a useful middle ground with the organization, scheduling, and maintenance chores taken over by the operator's office. The club then becomes a discount-rental arrangement with perhaps a designated meeting room and organized activities for family and social enjoyment. There is less personal effort required to sustain an FBO-sponsored club, because the sponsor takes over the dirty work.

The time would seem to be right for a rebirth of the flying club. All you need to do is surrender a little bit of your precious independence, working out an enriching partnership with a few fellow pilots that'll make it easier for everybody to fly. You could stand to gain much more than you lose.

FIRST LOVE

You have finally made a deal to buy an airplane after months of scanning the want ads, bugging airplane dealers, and hanging around airports. Oh, maybe it's not exactly brand new, but it's just what you've always wanted, and it's something you've been working toward for a long, long time. Congratulations; I hope you'll both be very happy.

But before you rush back home to drain your bank account, and forthwith return to take possession of this beauty, are you *really* sure you know what you're getting into? Oh, you do, huh? You say you had Joe the mechanic look the bird over and check its logbooks, following instructions that you read about. Good, good; now what about insurance, registration, and all the other stuff? You hadn't gotten around to those details, eh? Welcome to the ranks of aircraft owners, good buddy; you've got a lot to learn.

PAPERWORK

First, taking delivery of your new property involves a little paperwork, and some common sense as well. Your friend, Joe the mechanic, has seen that the required papers—such as the airworthiness certificate, weight and balance data, operating limitations, and logbooks—were all in order. Now, as the purchaser, you need to get the previous owner to sign a proper bill of sale, such as (but not necessarily) the FAA's AC Form 8050-2. Remember, if the aircraft was registered to a partnership, all partners must sign the bill of sale, even if you're buying only one partner's interest in the aircraft. No notarization is required.

You will apply for a certificate of registration on AC Form 8050-1, which must be sent with the bill of sale to the FAA Aircraft Registry in Oklahoma City, Oklahoma; the fee as of this writing is $5.00. The pink copy of the application is to be carried in the aircraft as temporary registration, valid for 120 days, pending receipt of your new registration.

It is important to realize that the FAA does not guarantee ownership; in other words, a registration certificate is not a "clear title" per se. Liens might be outstanding against the aircraft despite the apparent ownership conveyed by proper registration. Lien holders can register their liens with the FAA for another $5.00 fee, and before you purchase any aircraft it would be wise to have the FAA files searched for outstanding liens, a service performed not by the FAA but by private title search firms located in Oklahoma City. Such a search would be money well spent unless you are quite sure of the bird's history. There could, of course, be unrecorded liens not filed in the FAA registry.

If you have any plans to take your new airplane out of the territorial limits of the United States, you'll need to make application for a new radio station license in your name, issued by the Federal Communications Commission. There is no requirement for the FCC license if you fly only within the United States. In a similar vein, remember that the "pink slip" copy of the registration application, valid for 120 days, cannot be used to leave the U.S. borders; you must have the permanent registration in the aircraft.

LOCAL TAXES AND REGISTRATION

If you live in a state that features state registration of aircraft, don't forget to 'fess up to the local boys; they expect their tribute, usually on an annual basis. More important, check out the applicability of sales tax laws, both in the state of purchase and your own state, if different. Those sales tax investigators get pretty sharp about tracking down violators on big-ticket items like aircraft, and you don't want to get hit with a large penalty-laden bill later on. It'll cost more then, so you may as well pay up now. And you'll need to add the airplane to your list of personal property on your next property tax assessment.

INSURE THYSELF

It would be wise to avail yourself of some insurance coverage before you go winging away. Other owners can refer you to brokers or companies handling their insurance. The broker will need to know the airplane's identifying numbers and your qualifications in order to bind the coverage. This can be arranged before you go to take delivery, so you need not risk an uninsured trip.

While there is no federal law requiring you to carry any insurance at all, it would be pretty dumb not to have liability protection, at the very least. It's a hungry world out there, and people get sued by their friends every day. If you should inadvertently get involved in a mishap that kills or injures a passenger or an innocent bystander on the ground, you will need coverage. Protection against windstorm and other ground damage losses may cost relatively little additional premium. Also, at least some of the more expensive in-flight hull coverage may be needed to protect your banker's interest, if you have the airplane financed.

CHECKOUT

After seeing to the insurance, what about getting a checkout? If you're already experienced in that type of aircraft, you may feel qualified in the airplane after only a few minutes in the cockpit with the previous owner. Just make sure he tells you about all the knobs that don't work and the switches that don't switch. More than likely, however, you will need a qualified CFI to ride with you, one who is familiar with this make and model of airplane. A CFI checkout may also be a requirement of your insurance policy, without which you aren't covered.

If you are buying a high-performance airplane with over 200 horsepower, or a complex airplane with flaps, retractable gear, and a controllable propeller, you must have a certification of competence to fly such an airplane in your logbook. Once endorsed, you can legally fly any other type of high-performance or complex airplane. Be sure to mention this to the instructor who gives you your checkout, or he may not sign you off properly. There was an exception for old timers who had logged pilot in command (PIC) time in a comparable airplane before November 1, 1973; and before August 4, 1997, one endorsement covered both groupings, so if you recorded flights as PIC before then in both high-performance and complex airplanes you don't need the endorsements.

STAYING LEGAL

Any required ADs (airworthiness directives) to be performed on your airplane will be mailed to the previous owner until the paperwork changing the registration clears Oklahoma City, so ask him to forward such vital mailings. In case of multiple ownership, the first owner listed gets the mail. If a transponder is on board, make sure it's legal for use as required by regulation; whether VFR or IFR, it must be recertified by a repair station each 24 months. If you plan to fly instruments, you will need to check the logbooks for an altimeter/static system test within the past 2 years, and, of course, verify and sign off the VOR receivers as being within tolerance. And be certain you check the currency of the electronic databases in the GPS receivers or mapping devices.

As you can see, ownership means a lot of details suddenly become *your* responsibility. When an annual inspection is due, it's your obligation to see

Oh, wow—the check has cleared, the insurance is bound—it's time to fly away in your very first airplane.

that it's scheduled, not the mechanic's. When an AD is issued on your make and model, you are responsible for having it complied with, and are to maintain a record of such compliances. It would be well to have a little extra cash laid back for unexpected maintenance bills. In our enthusiasm, we often underestimate the cost of keeping an airplane, particularly an older, complex one such as a Bonanza or Cessna 182.

But there is a lot of satisfaction in pulling those keys out of your pocket and saying, "That's my airplane, I can go fly anytime I want; I don't have to wait in line or ask anybody's permission." Enjoy your new pride and joy, first-time owner!

CUSTOMIZED CHECKLIST

M any pilots believe the ability to fire up and take off without the aid of a printed checklist is living proof of their maturity. For them, it is an admission of weakness to resort to any sort of memory aid. Although it is true that your personal airplane eventually becomes so familiar that you seem to know standard procedures by heart, such familiarity can be a long time coming, and even then the stress of an unusual situation can make you forget. A simple list of things to be checked before, during, and after a flight can prevent many an interesting moment, and perhaps save a bent airplane.

Some need the checklist for other reasons. Rental pilots, as well as those owning an airplane in partnership, have to contend with another person's habits and fallibilities. If the last pilot turned the fuel valve off when parking, contrary to your habits, and you make a hasty departure without checking it, the engine could sputter to a halt at just the wrong moment. Were all radios left on by the previous pilot, ready to be injured by your start-up? Perhaps a fuse or lightbulb burned out on the last trip, and you are the next person to fly the unrepaired ship. By all means, if you share an airplane, have *and use* a checklist.

CHECK THE AIRPLANE, NOT THE LIST

One word of caution regarding the use of a checklist, however: **Do not place blind faith in a simple laminated card.** Familiarity can lead you to look at the list without actually reading it, and you may speed down the card, confident that you checked over the list, all the while skipping one or more items. You, as the pilot, are to check out the airplane, not the checklist. Move your finger or thumb down the page one line at a time, and touch or move the item checked with the other hand, using physical contact to confirm the check.

In general, checklists provided by the aircraft manufacturers are next to useless. The ones printed on the instrument panel or sidewall are too brief, and the lists contained in the pilot's handbook are awkward to locate and handle. The best checklists are still those on a single piece of durable card stock, small enough to fit the hand comfortably and big enough to be found easily among the litter in the map pockets or glove compartment. A 5- by 8-inch card is about right, although I prefer a slimmer 4- by 9-inch card, easily obtained by whacking a strip out of a file folder.

Don't acquire a new airplane and immediately print up a detailed, plasticized checklist for it. You will wind up wishing you had included extra items or arranged them in a different order. It's best to rough out a preliminary list, use it for several hours of flying, and make revisions to it until it fits your exact needs. You may not want a direct copy of someone else's checklist; every pilot has his or her special concerns or faults to be checked against. However, basic lists can be expanded to fit nearly any aircraft or pilot.

ARRANGING THE LIST

Divide the checklist into sections, which makes it easier to find where you should resume checking, by creating convenient "quittin' places." A "Before Boarding Aircraft" section might include switches off, fuel on, controls unlocked, preflight inspection okay, weight and balance checked, and so on. The next division, "Before Starting Engine," should be a logical sequence to check basic items and get the fire lit. "Seat Belts and Harness Fastened" is an item required by regulation, "Doors Shut and Latched" saves surprises later, and "Baggage Secure" keeps loose objects out of unwanted places. Brake pedals should be squeezed to test for an empty master cylinder. Controls should be checked for full, free, and correct travel. The fuel valve is reconfirmed "On" and all electrical and radio switches should be "Off," followed by normal starting procedure. This section should conclude with a few items to be noted after the engine is started, such as oil pressure coming up, primer locked, radios on, and anticollision beacon flashing.

You should now be ready to leave the ramp, letting the engine come up to operating temperature while en route to the runup area. When reaching the runup position, resume at the next section, headed "Before Takeoff." Again, controls should be checked, in case you picked up a cornstalk or something while en route to the pad, or just plain fired up without the checklist. Trim settings are important, and flaps should be cycled, whether manual or electric, to see if they work normally and at the same time. I like to spell out the setting of directional gyro and altimeter as specific items, but generally lump the scan of other gauges under "Check Instruments for Normal Indication."

Leaving the actual runup until near the end of the pretakeoff section gives the engine a few more seconds to warm before winding it up to the specified rpm for checking mags, carb heat, prop, ammeter, and vacuum, as applicable to your bird. Limits of mag drop and other parameters should be on the card; in case you get curious about an indication it saves digging out the manual. It is well to include another check of door security; it may have been propped open during taxi for ventilation.

I like to conclude the pretakeoff list with a recheck of the fuel selector on the correct position and a reminder to look for other traffic before taking the runway. You may have all sorts of extra items you want added to your airplane's list, so go right ahead; that's why you're making a list—to keep you or a friend who flies the airplane occasionally from letting something important slip by.

IN THE AIR

In-flight portions of the checklist should be kept brief, because the user will be busy with other duties and shouldn't be looking down at his or her lap too much. Best climb speeds and EGT-leaning parameters would be helpful reminders, and commonly used cruise-power settings can be included. I like to recheck "boost pump off" at the top of the climb, just in case I forgot to turn it off as I left the traffic pattern. The "Before Landing" list will include let-down items that may have been overlooked, such as mixture rich, cowl flaps open (or closed, as desired), and main tanks or boost pump selection. A gear-down item should be included early in the "Before Landing" section, and once again at its conclusion, to make sure a safe indication resulted from moving the gear handle. Gear and flap limit speeds should be on the list to help unfamiliar pilots who may miss the placards.

The "After Landing" items will include such reminders as carburetor heat off, cowl flaps open, flaps up, boost pump off, strobes extinguished, and transponder recycled to standby on code 1200. The last section, "Shutdown Procedure," should make sure all electrical items and avionics are turned off and a magneto grounding check needs to be performed before the mixture knob is pulled. In most aircraft, I find it convenient to follow a path from the radio stack or avionics master switch to the mixture knob, then across the electrical switches as the engine dies. The key should be removed from the magneto switch as a double-check that the switch is off, and the control lock should be installed. The final item should be "Secure Aircraft," a reminder to either tie down or hangar the bird before leaving it.

After revising your list into its final form, prepare to print out your checklist on good heavy card stock that will still feed through a printer; you may be able to get 90-pound index card or 11-point manila file folder material through the tray-bypass feed. Use a bold font so you will be able to read the list even in poor light. I like to have the entire pretakeoff list on one side of the card, so there is no need to flip it over to finish. Keep the wording simple so that use of the checklist will require a minimum of skill or deduction.

PRESERVING YOUR HANDIWORK

To protect the checklist, some sort of clear overlay should be used. A plastic envelope of the correct size would allow the list to be removed and updated, or you may prefer to laminate your efforts in plastic. The lamination adds stiffness and seals out grime as well; the sticky-back acetate sheets are available at most stationery stores. Cut the sheet a half-inch larger than your checklist to give a margin for sloppy placement. Peel off the backing and lay a sheet down sticky side up. Starting at one end, place the checklist on top of it, smoothing it down onto the laminating sheet gradually to keep air bubbles to a minimum. Small bubbles can be worked out by rubbing the surface with a blunt instrument. Get the card lined up right the first time, because there's no moving it once it makes contact! After both sides have been coated, trim the sealed edges of the plastic, leaving a quarter-inch or so of overlap to prevent delamination.

Put your name and aircraft number on the checklist, because if you do a good job of preparation it might be borrowed by someone who admires your work, never to return. Use the checklist; you'll find that it saves you time as well as brings order to your checkout of the airplane. It's just a simple reminder, but the times when you're in the biggest hurry are when you are most likely to skip a vital item.

SAMPLE CHECKLIST FOR CESSNA 182B SKYLANE

Before Entering Aircraft

1. Remove controls lock.
2. Confirm ignition off.
3. Check fuel selector "Both."
4. Perform preflight inspection.

Before Starting Engine

1. Weight and balance checked.
2. Seat belts and harness fastened.

3. Doors closed and latched.
4. Controls unlocked and free.
5. Brakes set.

STARTING PROCEDURE

1. Fuel selector "Both."
2. Mixture full rich.
3. Carburetor heat cold.
4. Cowl flaps open.
5. Propeller high-rpm.
6. Radios, lights, etc., off.
7. Master switch on.
8. Throttle cracked ½ inch.
9. Prime as required.
10. Propeller area CLEAR.
11. Magneto switch "Both."
12. Engage starter.
13. Check oil pressure.
14. Avionics "On" as required.
15. Idle 800 rpm for warm-up.

BEFORE TAKEOFF

1. Brakes set.
2. Controls free and correct.
3. Trims set for takeoff.
4. Altimeter set for field elevation.
5. Directional gyro set.
6. Instruments checked (green arcs).
7. Fuel adequate for flight, "Both On."
8. RUNUP—1,700 rpm:
   • Magnetos; max drop 125 rpm.
   • Propeller; drop in rpm.
   • Carburetor heat; drop in rpm.
   • Vacuum gauge; 4.6 to 5.4 Hg.

9. Acceleration test.
10. Flaps as required.
11. Recheck; prop in, carb heat cold, cowl flaps open.
12. Autopilot off.
13. Anticollision lights on.
14. Transponder on.
15. Clearance if required.
16. Check for traffic before takeoff.

CLIMB

1. Best rate 88 mph, best angle 70 mph.
2. Cruise-climb power, 23″ and 2,450 rpm.
3. Cowl flaps open as required.

CRUISE

1. Cowl flaps closed.
2. Normal cruise 22″ and 2,300 rpm.
3. Lean 100° rich of peak EGT.

BEFORE LANDING

1. Mixture full rich.
2. Fuel selector "Both."
3. Carburetor heat checked, on if required.
4. 12″ m.p. (manifold pressure) for approach.
5. Trim as required for 90 mph glide.
6. Autopilot off.
7. Flaps as required (100 mph maximum).
8. Propeller full in on short final.
9. Normal approach 80 mph, short field 70 mph.

AFTER LANDING

1. Raise flaps after rollout.
2. Carburetor heat cold.
3. Cowl flaps open.
4. Transponder and strobes off.

SHUTDOWN PROCEDURE

1. Check for ELT signals (121.5).
2. Engine idle.
3. Radios off.
4. Mixture to full lean.
5. Electrical switches off.
6. Magnetos off, key out.
7. Master switch off.
8. Fuel selector "Off."
9. Controls lock installed.
10. Secure aircraft.

TEST FLYING AN AIRPLANE

Most pilots will eventually encounter an airplane that has been taken apart and put back together again. Although it may not be officially required by regulation, any airplane that is fresh out of the shop must be given a test flight of some sort. It may be the first flight after a 100-hour inspection or annual, or it could be the aftermath of a major rebuilding job. Whatever the cause, if you're going to be the lucky person who gets to make such a flight, you should know how to go about it safely.

Test flying needn't mean a death-defying adventure with a parachute strapped on and crash trucks standing by; that's for *experimental* test flying, where the pilot is a flying engineer probing unexplored areas of a design's limitations. In most cases, the owner/pilot will be flying a tried-and-proven machine with only a little extra care needed to assure that all is well before returning it to full duty. However, it would be particularly wise to avoid night or IFR flight until the airplane has been shaken down, if you're picking it up fresh from the shop. If the task is somewhat experimental, such as when making the first flight of a rebuilt or a new homebuilt aircraft, don't be ashamed to seek a professional pilot, rather than flying it yourself if you're short on experience; there's no sense risking all that hard work.

RETURN-TO-SERVICE CHECK

Let's start with some of the more commonplace tasks—for instance, flying an airplane that has been given an annual inspection or thorough 100-hour inspection. In theory, such an airplane shouldn't need any extra care at all, because it has just been signed off as airworthy. People are human, though, and people make mistakes; some people are just more human than others. I have had shimmy-dampers left dangling, fuel caps left off, oil screens only finger-tight, and lots of other good things, all under the guise of routine maintenance. Of course, you're not going to trust your airplane to a shop you aren't sure of, but the FARs make it quite clear: The final responsibility for making sure an airplane is ready for flight rests with the pilot who is making that flight. So check everything, no matter *who* says it's ready to go.

If possible, spend some time in the shop along with the airplane, if you can manage not to make a nuisance of yourself. Wear old clothes, stay out of the way, and offer to handle screwdriver jobs like removing fairings or inspection covers. Look your airplane over while it's opened up; it's amazing how many things can happen inside over a year's time, and the more eyes that are in there looking, the more things will be caught instead of missed. The engine compartment requires about half the work of a light-plane inspection, so keep an eye out for chafing controls or cracked, worn, or loose items. Add anything found to the squawk list brought in with the airplane so it will come to the A&P's attention; don't interrupt his work every few minutes with trivia, or you'll have a very expensive inspection!

Keep a mental note of any repairs or alterations that could affect the coming flight, such as removal of control surfaces or partial teardown of the engine or accessories. Knowing what items have been repaired will give you a chance to monitor them closely for an unhealthy sign after takeoff. As the reassembly progresses, begin your preflight inspection; watch for wrenches hiding between cylinder heads, cowling screws not tightened, hubcaps left off, or battery cables disconnected.

If you are picking up an airplane that has been left at the shop, find the mechanic who did the work and learn what was done; don't depend on

When an airplane leaves the shop, it should be gone over with a fine-tooth comb. Never assume that a mechanic or helper can't make a mistake.

the itemized bill you picked up at the front office. The A&P may even want to ride with you to check the results of adjustments he made. In any event, preflight the bird with a fine-tooth comb, as if you've never seen it before. Visually check fuel and oil levels, control freedom, and the general security of fittings and covers.

Before starting the engine, check for full, unrestricted travel of throttle, propeller, mixture, carburetor heat, and cowl-flap controls, and see that flight controls move correctly in response to your wishes; aileron travel should be checked with the elevators in full-up and full-down positions. Start the engine, monitoring pressure and temperature indications, and warm-up no more than necessary for smooth acceleration to taxi power. Check the brakes as you begin to roll, before you get close to other aircraft or buildings; there should be even braking with no fade or odd noises. Check nosewheel or tailwheel steering during taxi, and park into the wind for the runup, particularly if the engine has been overhauled and is still "tight."

Check all engine instruments for normal readings; the fuel gauges should agree with your visual findings, and careful ignition and propeller-cycle tests made at the recommended rpm. If you believe a full-throttle runup is justified, keep it as short as possible and make sure you're not blasting people or property behind you. In most cases, you could determine the engine's full-throttle output at the start of the takeoff roll with plenty of room to stop.

As you make a normal takeoff and climbout, remain in the traffic pattern to gain altitude and note the aircraft's performance. By doing so you will be close to a landing spot if something suddenly goes wrong, rather than several miles away from the strip. Follow the shop's instructions about power settings and climb speeds with a newly reworked engine; be ready to increase climb speed or level off if overheating occurs. In most cases an airplane that has undergone no more than a routine inspection will exhibit few unusual traits, but be listening for new sounds or vibrations if you are familiar with the airplane's normal characteristics. I'll never forget the screws that were left untightened in an aileron inspection cover, falling out in short order until only one screw remained to hold the cover broadside to the wind; it took a half-deflection of the control wheel to maintain level flight until we got back on the ground. If you hear a new buzz or thump, land and check it out before you leave the area.

NEW OR REBUILT AIRPLANE CHECK

When **flying an airplane that has been rebuilt after extensive damage, or a spanking new homebuilt,** even more attention to preflight details is called for. Wait until winds favor the longest runway available, and do a little study of the forced-landing areas available nearby. If runway length permits, a partial takeoff will allow a sampling of control response just above the runway, followed by a landing straight ahead. Most standard designs are predictable if factory rigging is followed, thus producing few surprises for the test pilot.

In the eventual takeoff, allow the airplane to lift itself off well above stall speed to assure good control and check the takeoff trim setting for correctness as you accelerate into a climb. Waste no time getting to altitude, because this is your best insurance in any emergency, and avoid built-up areas and housing as much as possible. Make no power reductions until a healthy margin of altitude is gained.

If no unusual tendencies present themselves, continue a cruising climb to at least 3,000 feet AGL and level out to check rigging and stability. Trim should be adequate to allow hands-off flight in smooth air; observe any deviations from straight flight for ground adjustment later. If a turn consistently begins after the controls are released, is it the result of drag from a yaw of the nose or a true "heavy wing" that goes down before the heading changes? Which is needed to keep the skid/slip ball centered in straight-and-level flight, aileron or rudder? Bringing back such facts will tell the mechanic where to change rigging or add trim correction.

A correctly loaded airplane trimmed for cruise should recover hands-off promptly from a smooth pull-up to a speed 20 percent below cruise, by oscillating into diminishing climbs and dives until once again in level flight. Any prolonged climb or dive indicates poor CG placement or inadequate stabilizer area. Slight yaws induced with rudder application should snap back to straight flight when the pedal is released. However, most aircraft will stabilize into a spiral if a wing is permitted to stay down after displacement from level flight. A slow-flight check will sample control at near-landing speed and the amount of P-factor generated in a go-around.

Stalls should be conducted to calibrate the airspeed indicator before landing, by slowing at a rate of 1 mph per second in landing configuration until a definite stall occurs. A figure 1.6 times this speed should be maintained for safe maneuvering in a glide, and 1.4 times the stall speed used for final approach. If the stall occurred at 50 mph, 80 mph would be a safe traffic pattern speed and 70 mph a good approach speed for the first attempt. Any consistent wing drop at the stall break should be noted, but make sure you aren't inducing it by uncoordinated use of rudder during the entry.

Operate all equipment and controls during the flight, checking fuel feed from various tanks, electrical load carried as radios and lights turned on, and all flight instruments for normal readings. Pull on the cabin heater to check for exhaust smell as well as heat output; a fresh paint odor would be normal with a new engine, and a solvent aroma is to be expected if the engine has just been washed down.

If an engine is being broken in, make the traffic pattern under power to avoid excessive sudden cooling, A near-stalling touchdown should allow stopping without riding the brakes, although their stopping power should be tested briefly. Cool the engine down by a few minutes of idle or slow taxiing, and check that idle speed does not exceed 550 to 600 rpm with the throttle full back. Moving the mixture to idle-cutoff should produce an increase of 25 to 50 rpm before the engine dies, if the carburetor or fuel control is properly adjusted.

When you pick up an airplane from the shop, don't just ask if it's ready to go; before you blast off for home, spend a few minutes checking it out, no matter what anyone says. Be mentally prepared to be a test pilot for that first hour or so. The next hour might not be so convenient.

THOROUGH POSTFLIGHT INSPECTION

One of the things seemingly done decidedly backward in general aviation is total reliance on preflight inspections to assure an airworthy airplane. In reality, we ought to perform a systematic *postflight* inspection as well, so potential problems could be detected in plenty of time to get them fixed before the next flight, rather than left to resurface as nasty surprises after the next departure.

Typically, however, we roll to a stop, cut the switches, and swing out of the saddle, walking away with scarcely a backward glance, glad to have the trip behind us and going home. Maybe tomorrow we'll get back out to wash off the bugs, or perhaps we'll call the shop to put air in those low tires—if we don't forget.

In earlier days, when most of us flew for fun—or just for the novelty of it—we stayed around the airport after a flight to visit with other pilots while we tinkered with our airplane. Today is the age of the businessperson/pilot, urgently needed elsewhere as soon as the wheels stop turning. Airports are virtually devoid of puttering pilots; many of the current crop of aviators have never seen their airplanes with the cowling off.

And yet, when something turns up amiss on a preflight inspection, much wailing and gnashing of teeth accompanies reshuffling a schedule, while a feverish search is conducted for someone to render emergency service. About this time the pilot ought to ask herself, "Couldn't I have caught this on a postflight inspection, so it would be fixed now instead of broken?"

By noting items needing attention right after the airplane arrives, the maintenance crew can work on the squawks at their convenience, with less disruption in their work. Every mechanic likes to get one job done before starting on another; pull that person away from a deeply involved project to attend to a minor task—which really should have been taken care of as routine maintenance—and you definitely have not made a friend. If there is any way the lowest-ranking, least-trained member of the shop staff can be assigned to your problem, you can bet it will be done.

Okay, so a postflight inspection, followed up as needed by your mechanic, is a valuable ally in achieving smooth, hassle-free departures. By no means is a postflight check a substitute for the next preflight; preflight walkarounds still are needed, just as always, particularly if you share your airplane with other users, and most particularly if it isn't stored in a locked hangar. Verifying an airplane's suitability for flight just before taking off goes back to the earliest days of aviation.

SQUAWK SHEET

The emphasis of a postflight inspection should be on those things that could be fixed before the next flight, rather than on checking the aircraft's immediate airworthiness for the upcoming flight. It is important to keep a squawk sheet in the aircraft for postflight notations, both for your mechanic's benefit and to refresh your memory at the next preflight.

Start the postflight inspection while taxiing in after landing. Is brake action normal, pedal travel firm and even? With rudder trim centered and main gear struts settled down from their in-flight extension, is the airplane

Checking the oil before departing from home, this Mooney TLS owner makes sure his airplane will be flight-ready when he unties it next time.

rolling straight in the absence of a strong crosswind? The ammeter or load-meter should be showing no discharge or lack of load. If a long taxi run permits, bring the engine up to magneto-check power briefly and switch through the magneto positions to observe tachometer indications. Better yet, stop at the runup pad—if traffic permits—as you exit the runway for a stationary magneto check. Do not perform runups in a tie-down space or in a hangar alley, where the prop blast might throw debris into other air-planes and make excessive noise.

Before pulling the mixture, turn the magneto switch off for a split second while the engine is idling; the engine should have stopped running while the key was in "Off." This allows you to see if you've lost a grounding lead on the flight, which would render a magneto "hot" even with the switch off. If this is the case, placard the propeller "Do not touch—live magneto."

Also, tune the ELT's emergency VHF frequency, 121.5, on a com radio, turn up the volume to check for any ELT wails, then go back to the local channel. Reset the transponder to 1200 at this time and tune the nav radios to the local nav stations. All avionics should be turned off before engine shutdown to avoid leaving one on during start-up when voltage spikes come through the wires.

Then, pull the mixture into the idle cutoff and watch the tachometer for a 25-rpm rise just before the engine dies. The absence of this rise denotes an idle mixture setting that is too lean; more than 25 rpm means it

Once tied down, a few minutes spent walking around the airplane before leaving the airport can spot items to be corrected prior to the next flight, like a low tire or missing screw.

is too rich. Either should be adjusted when other routine work is performed under the cowling.

AFTER SHUTDOWN

Note the tachometer time and compare it with the last oil change and 100-hour inspection time; better to discover you're needing the shop's attention now, rather than when you're already about to get underway for another trip. If you're the studious type, start filling out a trip sheet after each flight; time en route, fuel added, oil consumed, any pertinent instrument readings, temperature and weather conditions, along with anything else that might help analyze trends and improve aircraft operation.

Outside, **give the airplane a quick inspection for fresh oil leaks,** looking under the belly, beneath the cowling if possible, or, at the very least, observing any streaks on the sides of the cowling. Take an oil reading as soon as the dipstick has cooled enough to be handled, bearing in mind that some oil might be retained in the galleries until 15 minutes have elapsed. If it is low, bring the oil level back up to a normal operating range, so you won't have to hunt a quart of oil during the preflight. Clean off any oil drips or streaks while the warm oil can be wiped away easily, rather

A slipping starter could have been brought to the attention of the A&P mechanic when the flight was over, instead of after it failed completely. Make notes of such "squawks" so you won't forget them later.

than waiting for it to set up and attract dirt; you can prolong the intervals between solvent washdowns by doing some postflight wiping.

This also is an excellent time to clean the windshield, something that often is neglected in the heat of departure. Bugs can be removed easier when they haven't had time to dry, so you might want to sponge off the aircraft leading edges while you're at it, depending on the time of year. After operating at a grass strip, grass clippings should be brushed away before they stain white paint or dry into cementlike slurry around struts and brakes.

The landing gear should be examined for looseness, low strut extension, or accumulated dirt and mud. Most tires will need air periodically, and maintaining proper inflation extends tread life greatly. If wheel fairings are installed, roll the airplane to check all sides of the tread, noting any flat spots from brake lockups, cuts, exposed cord, and wear patterns. Spring-gear Cessnas require main landing gear tire reversal at the midpoint of the tire's lifespan because the wheels droop in flight and the outside edge of the tread always contacts the runway first. Neglecting to have this done results in throwing away half the tire tread when cord starts showing up.

DAMAGE DETERMINATION

Look over the exterior of a sheet metal airplane to check for loose or missing fasteners; loose cam-lock fasteners leave dirty-appearing circles of aluminum oxide around their perimeters. Carry some replacement PK screws so you can install them on the spot if any are missing. Check for any new dents or dings that might have resulted from the flight; anything serious enough to require a mechanic's inspection goes on the squawk sheet. Control surfaces are rechecked for smooth travel, hinge tightness, and lack of damage.

This is a good time to remove the battery's cover and check the electrolyte level, particularly if you're sporting one of the little 24-volt batteries with 12 cells. A hardworking battery uses electrolyte, especially toward the end of its life, and if you allow the plates to become uncovered you're hastening the battery's demise. Don't figure on letting the mechanic catch it at the 100-hour inspection, because tired batteries won't run that long without a refill. Use distilled water, of course.

Turn on the aircraft lights during a postflight inspection. Now is the time to install bulbs, not when you're trying to go somewhere in a hurry. Also, the airplane will be ready for a night flight when you want it to be, if you check it regularly. Replace any missing fuses that belong in the spare holders if you're not equipped with circuit breakers. See if the clock is running; the 1-amp fuse feeding its circuit might have blown.

If the airplane is going to be left outside, close all the vent openings to keep out rain and install sunshields if you have them. If not, spread out a bedsheet or some charts to at least provide a little shade. Control locks are to be installed, of course, with external gust locks added for longer-term outside storage. Install the pitot cover, if available. Most savvy owners also have various forms of pest control to be installed; bird stoppers for the cowl openings, plugs for the tail gaps, mice deflectors around the landing gear, and so forth.

Tie-down ropes or chains should be tied as snuggly as possible, the only exception being new hemp ropes, which can shrink after taking on water and drying out. Chock the wheels, rather than set the parking brake, to avoid overpressure in the brake lines as the hydraulic fluid heats up and expands, or bleeding off if the temperature drops after the brakes were set. Spring-loaded flaps should be locked down, so hinges won't be beaten back and forth in the wind.

Now that you've taken care of the airplane, go see your mechanic about any items requiring attention before the next flight. Then you can head home with a good feeling about that next preflight inspection, which should hold fewer nasty surprises.

ACCIDENTS

Despite all your precautions and every last-ditch effort to prevent it, your turn comes: you have an accident, you break an airplane. Accidents are always supposed to happen to the other guy, but suddenly you've become the other guy.

When the crunch comes, your first priority is always preservation of life and limb. We can replace airplanes, but when people are killed or hurt they can never be made whole again, as if it had never happened. After the crash trucks are gone and the dust settles, the brief paragraphs of National Transportation Safety Board Regulation Part 830 take on a new significance. There are procedures to be followed and regulations to be satisfied.

ACCIDENT DEFINED

What constitutes an accident, anyway? Contrary to popular opinion, it is not based on a dollar estimate or a personal injury index. Rather, the National Transportation Safety Board (NTSB) calls an aircraft mishap an accident, for reporting purposes, when substantial damage is done to the aircraft, or serious injury occurs to a person resulting from the aircraft's operation. *Serious injury* means immediate or subsequent hospitalization for injuries suffered in the accident. A *fatality*, of course, always makes it a reportable accident, even if the aircraft is not damaged. Also, damages in excess of $25,000 to property other than the aircraft will make the event reportable.

Substantial damage would be any damage rendering the aircraft unairworthy, such as bent wing structure, crumpled fuselage, broken tail surfaces, or a damaged firewall. Nonreportable damage would be that which is limited to the engine, propeller, landing gear, flaps, skin, or wingtip.

Therefore, you can have a connecting rod failure, land in a rock-strewn pasture where you blow out a tire and puncture the fuselage fabric, and hit your nose on the panel hard enough to make it bleed profusely, and you still have not had an accident, by the terms of the NTSB regulations. Yet if you veer off the runway and run into a hangar, striking the airplane's wingtip hard enough to wrinkle the wing spar, it would have to be termed significant damage and reported.

NOTIFICATION, REPORTING, AND PRESERVATION

The time interval for reporting an accident is often misunderstood as being 10 days or 72 hours; actually, the NTSB must be notified *immediately*, using the telephone number found under U.S. government listings in the phone book. Remember, your obligation is to the NTSB, not the FAA, which has its own agenda. This initial phone call is usually followed by a written report on an NTSB form, which is due in 10 days. If you're unfortunate enough to be involved in an accident, there are certain basic procedures that should be followed. The owner of the aircraft should be notified at once, if not on the scene.

Enlist aid to secure the accident site from scavengers; the general public has a ghoulish fascination for wrecks and will drive for miles just for the chance to pick up souvenirs, important pieces that could be the only clue to the accident's cause. Even if they leave empty-handed, their footprints can track up skid marks and impact areas important to the investigation. Law enforcement officials are generally familiar with the site-preservation aspect of investigation, so call the gendarmes at once. When working with a wreck be alert for hazards, such as broken glass or plastic, open accumulations of fuel, and torn sheet metal. If it is necessary to move any part of the wreckage for safety reasons, or to change the position of a valve or control, mark the object carefully to guide the investigation team. Check to see if the emergency locator transmitter has been triggered and if so, turn it off, tagging the unit "activated by crash, turned off."

The more serious the accident, the more involved the investigation becomes. If no injuries result, and the circumstances were clear-cut, an investigation might not be made at all and the entire matter may be handled over the phone or by mail. But if injuries or fatalities occur, an NTSB team or FAA personnel will be dispatched to the scene, with the FAA usually handling the less-serious, non-air-carrier accidents.

Remember, do not move the wreckage until cleared to do so by the investigating authority. It would be wise to shoot some pictures of the scene in case questions arise later. As soon as possible, the wreckage should be moved to an out-of-sight area to lessen the crowd attractions. A locked hangar is best; failing this, a sheltered spot out of sight of the general public should be used. The pile of debris may be sitting for some weeks before a final insurance settlement and disposition is made, so plan accordingly.

INSURANCE

The aircraft owner would want to notify her insurance underwriters as soon as possible and, if the pilot is not the owner, she will soon learn about the coverage she didn't have. A rented or borrowed aircraft is not necessarily covered against mishaps while in the hands of a nonowner pilot and, even if covered, subrogation may be undertaken to recover damages from the pilot. For this reason, the pilot should make no statements that could be taken as an admission of liability unless she understands the consequences. On the other hand, it is helpful to write down what occurred as soon as possible to prevent forgetting some vital detail that could help explain the sequence of events.

Forms will be forthcoming—in quantity. Make copies of everything for your own files. The NTSB will want Form 6120.1 or 7120.2 completed within 10 days; it simply lists the circumstances of the accident and gives a statistical breakout of the pilot's qualifications and experience. The insurance company will have a similar form, parts of which may require completion by both the pilot and the owner, if different. If it appears that the pilot may have precipitated the accident by poor flying technique, the FAA can issue a request that the pilot take some dual instruction in the areas deemed deficient, such as crosswind landings, fuel management, instrument procedures, and so on. This form is to be signed off by an instructor

All occupants safely escaped from this wrecked Skyhawk; now the paperwork and legal maneuverings begin.

after the required dual is completed, although a further flight check by an FAA inspector may be needed to return the pilot to flight status.

Every accident should be treated as a learning opportunity. Bad as they are, they can serve a purpose in educating the pilot involved and all others on the scene. Unfortunately, experience is a very expensive method of learning and we must strive to prevent each mishap from occurring, to hold down our insurance rates and present a more positive image to the general public. But if an accident does happen, make the best of the situation and learn something from it.

GOING ON

FLIGHT REVIEW DUE?

It will usually happen once a week; a fellow pilot will seek me out to ask in a quavering voice, "How rough are you gonna be on my flight review?" As a flight instructor, empowered by the eternal and the FAA to make a pilot airworthy with one stroke of a pen, I am expected to be a cross between a fire-breathing dragon and a father-confessor. Shucks, fellas, I was drafted into this outfit just like you were; I'm as anxious as anybody to make it through without complications.

The flight review is like a misbegotten relative nobody wants to claim. Its origin and early childhood weren't widely discussed, and because fiction often makes more enjoyable conversation than fact, the reputation of the flight review remains besmirched. Like the sailor's tattoo of a green alligator, the flight review seemed to be a good idea at the time. But when the proposal to require an instructor's evaluation every 24 months was cast into a regulation back in 1973, nobody knew what to expect.

The instructors weren't told much other than to look for proficiency commensurate with the ratings held, and the pilots were simply told to get an instructor's signature in their logbooks or quit flying. With this lack of guidelines, pilots didn't know what to prepare for and instructors didn't know how to conduct the ride, and the first round of flight reviews varied widely in content and complexity. Fortunately, industry and government got together after a year or so and devised some nonregulatory guidelines to help straighten out the mess, but the bad taste from the initial poor start stayed with us for a long time.

The lack of enthusiasm for flight reviews also probably stems from their manner of promulgation. They were more or less demanded by the National Transportation Safety Board as a way of reducing pilot-error accidents, and everybody consulted agreed that it sounded like a good idea. But the FAA wrote only the basic law, at NTSB's behest, leaving the rest up to nature. Meanwhile, pilots thought they were doing quite all right, thank you, and suddenly here was another rule thrust upon them against their wishes. "Here, take it, it's good for you." Blah! You get the idea. To get cooperation, you must first convince the patient he's sick.

Because the word "biennial" appears nowhere in the regulation, we might have suspected that the FAA was planning to increase the frequency of flight reviews. Sure enough, as of September 1989 VFR pilots with fewer than 400 hours of total time were told they were going to have to take a flight review every 12 months, but that rule never took effect because there was no basis found for it. A provision requiring a minimum of 1 hour of ground instruction and 1 hour of dual was written into the law in 1993; the review has been unchanged since then. Future changes might include flight reviews for each category and class of aircraft for which the pilot is rated, rather than just one of his choosing, and perhaps even type-specific reviews for complex aircraft.

MAKE THE BEST OF IT

It behooves you to make the best of the situation; **as long as you have to have a flight review anyway, make it something productive.** Ideally, a pilot could get a new rating of some sort every 2 years and save the need for a flight review altogether, because any flight check takes the place of the review. But most people aren't interested in collecting ratings, and even so there is a limit to the ratings one can afford to acquire. Short of pursuing a new rating, one could schedule a good round of dual instruction and avail oneself of some instrument refresher work, aerobatic dual, mountain experience, or whatever.

Making one last check of his airplane and charts, this pilot awaits his flight review, a biennial endorsement by a CFI required of all pilots.

So maybe you're not crazy about the idea of entrusting your soul to an instructor after 2 years of freedom? Perhaps you're afraid he or she will uncover some hidden flaw that'll wound your pride? Or are you just "too busy"? Whatever your reason for avoiding the flight review, *don't*. Sure, nobody's likely to check you, especially if you're only flying your own airplane. But if you should have the misfortune to be involved in an accident while operating without a valid endorsement, you're going to be in a heap of trouble.

Oh, you can bluff your way past the FAA; about all they can do is revoke or suspend your license. But when your insurance company finds you were in violation of the FARs, you could get hurt in the pocketbook because they might just refuse to settle your claim. So, if you aren't going to take a flight review, you may just as well cancel your insurance and save the premium money.

Fear not; there is surely an instructor somewhere you can take into your confidence, swear to secrecy, and fly with long enough to prove your prowess. Looking at it another way, if you're good enough to "pass" the review, what are you worrying about? And if you really are doubtful about your chances of being good enough, you *need* the review. I can tell you that some sort of recurrent training is needed, though. We've always preached that, but only the International Flying Farmers organization ever believed in it strongly enough to make it a requirement as part of their discount insurance coverage. So we got the flight review as a result of our intransigence.

It's amazing how many nervous pilots I ride with on flight reviews. Even if I tell them they're going to pass unless they scare me, they still sweat out the ride with me, a harmless old "fright" instructor. I couldn't bust anybody if I wanted to. About all I can do is hit the ground running after the ride and refuse to sign your logbook, if you're really *that* dangerous. So get your favorite CFI to tell you what to expect and find out how you stack up.

TYPICAL PROBLEMS

As an example of the lack of self-initiated practice, consider the discussions I get into about stalls. I have ridden with numerous pilots who haven't stalled their airplane since the last flight review. A couple of stalls every 2 years is no way to maintain your mastery over the machinery, friends, and it shows. If your airplane is so dangerous that you're afraid to practice some stalls at altitude every month or so, you shouldn't be flying it. By the same token, if your proficiency is so lacking that you don't trust yourself to do a simple stall, what makes you think you're sharp enough to make a good crosswind landing or a short-field takeoff?

Another sore nerve I frequently touch during a flight review is the forced landing. Lots of pilots haven't heard the prop windmilling since their primary training days; every landing they make requires a lengthy, power-boosted approach, aimed at a mile-long ribbon of concrete. They have no idea of their airplane's sink rate or glide ratio with the power off because they haven't practiced a forced landing since they flew a light trainer. I can give you the names of a few pilots who are darned glad they had practiced their engine-out technique; the ability to execute a forced landing saved their bacon when the fan suddenly stopped turning one day. Unfortunately, I can also count a couple of friends who aren't around anymore because they weren't up to handling a sudden power failure. So don't wait for the flight review to come around to shoot a landing with the engine at idle; give it a try periodically.

Know thy airplane. This is a basic tenet of flying we often overlook, a fact that usually shows up on the flight review. What is your airplane's best angle of climb airspeed? You really ought to know in case you need to use it someday. What's your bird's useful load? Don't give me that old saw about "She'll carry anything you can shut the door on"; if I hear that one again I'll choke. I want the flight review applicant to know the airplane's numbers, the basic data every pilot should have down cold before he takes the airplane up. Yet I have reviewed pilots who have owned their airplane for years without learning the operating details. If you can't tell me how to safely jump-start your airplane, or what your fuel system's weaknesses are, or what your rpm and manifold pressure gauges should be reading as you roll for takeoff, you're going to know before we part.

BE CHOOSY

Don't ride with just anybody who happens to possess an instructor's certificate. Find an instructor whom you feel can teach you something and get some good out of fulfilling an obligation. If you're an instrument pilot, find a practicing CFII to give you the review. If you fly multiengine airplanes, you need a many-motor instructor beside you on the flight review. If you are a glider guider, hunt up an FIG (flight instructor, glider) for assistance. If you're an airline transport pilot (ATP), get a fellow graybeard to ride with you. If you're an antiquer, homebuilder, or warbird freak, ride with one of your own kind. Don't waste your time with an instructor who can't relate to your type of flying.

All instructors—myself included—occasionally forget what we're about on the review ride. The flight review is not a period of dual, it's an evaluation of a pilot's knowledge and ability. The dual can come later, if needed or requested, and we instructors should try to avoid giving lectures or demonstrations during the flight review. I am not supposed to remake every pilot into my image of the perfect aviator; my job is just to see if the specimen sitting beside me exhibits any deviate tendencies that could bring him to grief if not corrected. As long as the flying mistakes and blank responses are minor enough to be covered with practice and study, the flight review can be signed off with a little dual recommended. If el piloto is just an accident going somewhere to happen, however, the dual will be required before I can sign him off.

The success of the flight review concept depends utterly on the general public's response to it. Treat it as an opportunity, make it work for you, and you'll benefit from it. Ignore it as needless harassment, and you'll eventually prove the need for just such a requirement. The choice is yours.

AEROBATICS

Joining the ranks of aerobatic pilots is an excellent way to expand your flying horizons beyond basic point-A-to-point-B aviating. At this time there is no pilot rating for aerobatic flight; if you would like to try upside-down flying, the way you go about it is up to you. Unfortunately, many pilots attempt aerobatic maneuvers without fully understanding the dangers involved. It looked so easy when the air show pilot did that loop or barrel roll, yet one must realize there is a fine line drawn between safety and disaster for the unknowledgeable pilot.

Airplanes aren't made of boilerplate; they are built only as strong as necessary to meet certain standards of certification. Otherwise, they would have almost no payload and would be of little practical use. Naturally, an aerobatic airplane must be designed with excess structural strength as well as superb control ability in order to withstand the stresses of a maneuver that doesn't work out as planned. In addition, a good aerobatic ship is also inherently a little unstable—not always a desirable characteristic in every-day flying.

DON'T OVERSTRESS YOURSELF

As an example of the dangers involved in simple aerobatics, it is possible to fly an aileron roll within a maximum G loading of 1.5 to 2 Gs, well below the stress limits of even normal and utility category airplanes. Yet if the pilot errs slightly in nose position as the airplane goes inverted, the maneuver can turn into a sloppy rolling split-S, and more than 4 Gs easily can be pulled in the resulting recovery. Throw in a little rough air, or panicky control handling, and the airplane could be severely damaged from overstress.

True, a utility category airplane has a 50 percent overload factor, beyond its design load factor of 4.4 positive Gs, making the ultimate yield point 6.6 Gs, but this was for a factory-new, unfatigued, uncorroded air-frame. Also, to exceed the design load factor even slightly can still result in permanent deformation of the structure, even though nothing actually falls apart. The partial failure then lies in wait for the next pilot, who could get

225

A nice aerobatic biplane on a beautiful day is sure to bring a smile to any pilot's face.

into some rough air in the course of a normal flight and experience structural failure, all because somebody tried to prove his or her prowess as a hot-rock aerobatic pilot.

LEARN AEROBATICS THE RIGHT WAY...

As in all flying, there is a right way and a wrong way to learn about aerobatics. The right way is to use a tested, proven airplane designed for aerobatics with the guidance of an experienced aerobatic pilot; if he or she is a CFI, so much the better. Bear in mind that even certificated aerobatic airplanes have their limitations, and can't be tossed about in any maneuver you choose at any entry speed you like.

Pilots interested in trying out new maneuvers or procedures will wear parachutes and seek high altitudes over unpopulated areas for their experimentation; that's one reason cabin-type aerobatic machines are equipped with quick-jettison doors. Only when operating within the certificated limitations of the airplane can a reasonable level of safety be achieved; if the book doesn't show approval for outside loops, don't try 'em!

An accelerometer, or G-meter, should be on board, both as a warning against high stresses and as an aid to precision flying. You cannot always feel G-loading well enough to stay within limits, especially as tolerance to aerobatics is acquired. However, aerobatic novices will probably not watch the accelerometer or airspeed indicator during their first snap rolls or spins— they're doing well just to stay oriented! A shoulder harness—or even better, a five-strap restraint system with a backup safety belt—is vital to hold you firmly in your seat during all attitudes.

Again I stress, *don't try to teach yourself aerobatics from scratch,* no matter how easy it looked when you saw it done, or how simple it sounded when described. Get some time with an experienced, knowledgeable pilot; there is no substitute for hands-on flying through a maneuver with a skilled person at your side.

...OR LEARN AEROBATICS THE WRONG WAY

Most important, don't use a stock airplane for aerobatics. A standard Cessna 152 is a different bird from a 152 Aerobat, and it simply isn't strong enough to withstand the possible G-load resulting from mistakes. Sure, you've seen some pretty flashy aerobatic routines performed in stock airplanes at air shows, but these were done by a pilot who had learned aerobatics first in a beefy airplane, and who flew the stock airplane very carefully to avoid exceeding its lower stress limits. We all know a utility-category Shrike Commander can do slow rolls and loops; the trick is to do them safely, within limits, and without getting into a low-level trap with no escape.

You might have heard that older airplanes, such as Luscombes or Taylorcrafts, are okay for aerobatic training, because they do not have "aerobatics prohibited" in their limitations and "they built 'em tough back then." Yes and no; these old birds were licensed without specific limitations on specific maneuvers, leaving it up to the pilot to decide what to attempt. Today's airplanes are certificated only for specific maneuvers at certain maximum entry speeds, giving the pilot more guidance with which to stay out of trouble. In any event, bear in mind that a 50-year-old airplane may have brittle bones; the structure should be closely watched and G-loads kept light.

A good aerobatic flight is not a wild dogfight in the sky, tumbling through one maneuver into another, helter-skelter. Planning is needed to ensure that available altitude is expended wisely, and that maximum use is made of airspeed and position to keep the sequence flowing smoothly. Sure, free-style aerobatics is great stuff, but not for training, and not without a *thorough* background in basic aerobatics first.

It is not my purpose here to write a primer on aerobatic flying; you should invest in one or more good manuals already available for that. I am interested in making sure you are aware of the need for such manuals, as well as why aerobatic instruction is important. In most standard airplanes, airspeed control is rather critical and the unusual attitudes of aerobatic flying make this even more important. A clean airplane, such as a Bonanza, simply cannot be left nose-down for more than a few seconds; the airspeed quickly gets into the yellow arc and heads for the redline, where disaster awaits the unknowledgeable yanking on the controls. Should V_{ne} (redline airspeed) be exceeded substantially, a fatal control surface flutter might develop; loss of a control surface usually makes a recovery impossible.

Learning aerobatics involves knowing that such dangers exist and how to combat them. If airspeed is already plentiful, a nose-down attitude calls for the throttle to be cut to idle *immediately* to lessen speed buildup. If the nose is nearing a vertical attitude, on the other hand, and a tailslide is not desired, some thought has to be given to making a controlled exit from the straight-up position before all flying speed is lost. If a maneuver involves inverted flight, somebody is going to have to keep the nose up and roll the airplane back into normal flight without letting the airplane take matters into its own hands. Most important, spins need to be practiced so they can be recognized and stopped when they occur inadvertently, such as when attempting to half-roll off the top of a loop.

Learning aerobatics is a fine challenge, a great feeling of having a one-ness with the aircraft and true mastery of anything the equipment can do. But it is also a skill to be acquired properly, with full awareness of the dangers involved.

TAMING A TAILWHEEL

As the decade of the 40s drew to a close, the assortment of airplanes at a typical airport looked quite a bit different from those seen today. The great majority of aircraft were taildraggers, so much so that tailwheel landing gear was then termed "conventional" gear. Flying a taildragger in those days was not considered the mark of a hairy-chested individualist, merely part of the accepted order of things, just as was strapping a parachute on a student pilot to practice spins before being allowed to solo. You learned to fly with a tailwheel, and when you bought a new airplane it would also probably have a tailwheel. Actually, the "hot" ships of the day were the ones with nosewheels: Bonanzas and Navions. They and a mildly popular pre-war design, the Ercoupe, made up most of the tricycle gear population.

Out in Wichita, a small lightplane manufacturer by the name of Cessna was offering a complete line of three airplanes, all with tailwheels—the two-place 140, the four-place 170, and the ultimate luxury ship, the 195. There were speculations of a new design, perhaps to be called the 180, which was to be introduced as part of the Fiftieth Anniversary of Flight in 1953. Naturally, it would be a taildragger. Meanwhile, Bill Piper's bustling Lock Haven factory was turning out Cubs and Pacers, but some folks talked about a new Piper they had seen being tested, a Pacer with a nosewheel instead of a tailwheel.

The rest is history. Piper brought out the Tri-Pacer in 1951, and it sold like hotcakes as soon as pilots found out what they'd been missing. Cessna saw the light, and in late 1955 the 172 was born, followed closely by the 182. In successive years, tailwheel airplanes have shrunk in numbers, and many wild tales have grown from the mystique surrounding the older designs. New fliers are often told "You're not really a pilot until you can handle one of these babies," and nosegear-trained pilots who sampled a taildragger found that they did have a lot to learn.

"Why the tailwheel?" asks the novice pilot. When hard pressed, most taildragger pilots find it difficult to answer. Sure, less weight and drag, better prop clearance on the ground, less trouble in the bush country, all are cited as reasons, but few pilots flying taildraggers actually need these slight advantages, so it boils down to a simple preference. The same kind of person who likes to shift her own gears in her car, or control the exposure settings on her camera, will like a tailwheel airplane simply for the joy of having mastered it. Pilots are like that, and the pleasure of doing something for no real reason is one of the joys of living.

TRANSITIONING

FAR 61.31(i) requires that a pilot obtain a logbook endorsement from a CFI attesting to her competency to fly tailwheel airplanes, unless pilot-in-command time in such an airplane was logged before April 15, 1991. How difficult is the transition to tailwheel flying for the pilot who has never flown anything but nosegear airplanes? If she was taught properly and has kept increasing her skill, the transition will be relatively simple. For the not-so-sharp pilot, it will be thrilling—both for her and her instructor. I usually expect a 10-hour transition program if the pilot has never flown tailwheel airplanes.

As with most new subjects, a little preflight study of basic theory helps hasten the in-flight learning. A tailwheel landing gear is arranged with the center of gravity of the airplane behind the main gear, so that enough weight rests on the tailwheel to hold it down and provide steering friction.

The classic Beech Staggerwing flies just like any other high-performance airplane in the air; on the ground, however, the pilot had better be prepared for tailwheel characteristics.

The tricycle-geared airplane, on the other hand, has the CG located in front of the main gear, and the nosewheel supports a portion of the weight when at rest. This placement of the CG is an important point, because it affects the stability of the airplane when it is in motion on the ground. With the nosegear arrangement, the CG acts to pull the airplane back to a straight path if the longitudinal axis becomes misaligned with the ground track. But the taildragger's CG, which is "pushing" the main gear, adds exactly the wrong force when displaced, tending to get the tail around in front of the main gear. If not controlled, this is the start of the famous "groundloop" so colorfully embellished in hangar flying tales.

In a classic groundloop, the tail swings around in a tight circle, the inside wing rises and perhaps the inside main tire leaves the ground, and the tail comes up as the airflow no longer passes over the elevators and the panic-stricken pilot applies the brakes. The airplane can come to rest inverted in extreme cases, usually without injury to the occupants until they unfasten their seat belts.

To stop the groundloop before it goes too far, push full opposite rudder, chop any throttle, use only the outside brake, apply full aileron to the inside of the turn, and pin the stick full back.

RESPECT!

Simply put, the tailwheel airplane has limitations, just as do all airplanes, and the pilot must respect them. You must not allow a slight swerve to go unchecked, as you might with a nosewheel. Instead, constant attention to maintaining a straight ground path is required. Small corrections, made early in the game, are needed to prevent a groundloop from ever beginning. The poor ability of the small castering tailwheel to resist a crosswind component and the generally slower touchdown speeds of the taildragger combine to mean somewhat less crosswind can be handled by the inexperienced pilot than with tricycle gear.

Finally, control of the ground handling situation requires that the tailwheel be in constant, firm contact with the surface; otherwise, the main wheels will be free to choose their own path to travel. Placement of the elevators to maintain this contact is important, and most of the time the stick should be held full back so the elevators will be up to catch propwash and headwind.

To begin with, the pilot of a taildragger must have her seat properly adjusted. She should be able to push full travel with the rudder pedals without stretching, yet not be so close as to stab a toe brake accidentally. Land a nosewheel airplane with brakes applied and you merely stop quickly in a cloud of expensive smoke; land a taildragger with brakes on and you will stop even quicker—inverted! Some older designs have heel-actuated brakes, a real pain because the rudder bar moves but the brake pedal doesn't, and

This Piper J3C Cub, as well as the Taylorcraft and Cessna 180 behind it, are examples of how airplanes used to be built. The aft CG configuration means touchy ground handling for the uninitiated, requiring a proper checkout.

at full travel it's difficult to reach both rudder and brake. But the arrangement does have the advantage of keeping the pilot away from applying the brakes unintentionally.

TAXI, THEN FLY

During the first attempt at taxiing, you might feel that you are not in full command of the airplane, and probably you aren't. That's why you have a CFI with you. Although a large pneumatic tailwheel tire and properly adjusted steering gear will give almost as much directional control as a spring-steered nosewheel, the positive, crisp control just isn't there. Stay with the airplane, using early rudder application to keep the direction desired, and try to use less and less control travel as you gradually catch up with the airplane. As a last resort, use the brakes to stop a swerve that has gotten out of control, but don't steer with brakes exclusively. Remember, keep that stick back to hold the tail down, especially when the brakes are being applied.

A certain amount of forward visibility on the ground is sacrificed with the tailwheel configuration, so you may need to peer around the cowling and S-turn during taxi to clear the path ahead. Enjoy it. That bit about the steely eyed birdman leaning from the cockpit is part of the taildragger mystique.

Taxi slowly—no faster than a running walk—and when taxiing downwind move the stick forward, deflecting the elevators downward so the wind will hold the tail down. Park headed into the wind for the runup, so the wind and propwash are both available to hold the tail down. Keep the stick back during the runup, and don't let the airplane creep, because abrupt braking lightens the tail.

Checklist completed, it's time to have a go at high-speed taxiing. Use the runway (or a wide taxiway) for some mock takeoffs and landings, running up to near liftoff speed then cutting the throttle and slowing to a stop. As the throttle is fully opened and speed picks up, the tail will rise and the rudder, rather than the tailwheel, will play an increasingly important role in controlling the airplane. The tail can be held down for positive steering at first, but you may find it more helpful to get the tail up quickly for a better view of the runway ahead, which will aid your perception of misalignment. Don't worry if the instructor has to bail you out of some wild swerves at first; that's the way everyone began in the early days. Remember, use small corrections to catch errors early. Less and less rudder travel will be required as speed picks up. Constantly "walking," or fanning the rudder, was once taught as a coordination device, but it should not be necessary as soon as you get the feel of the airplane.

The tailwheel airplane is not finished flying until it is brought to a complete stop, much more so than a nosegear airplane, so don't go to sleep in the final stages of the rollout. Some large rudder displacements may be necessary at the last, aided by a touch of brake. Don't try to turn before slowing to taxi speed, or you may induce a swerve that is beyond the tail-wheel and brakes' ability to stop.

Having been emboldened by the high-speed taxi runs, you are now game for the full takeoff and landing, which is straightforward enough after the aforementioned preparation. Apply full throttle, hold a very straight track as the tail comes up, and keep the tail low by adding back pressure as speed picks up. The airplane will "float" off the runway at a much slower speed this way, and naturally you could let the tail come up higher for a faster liftoff speed in gusty crosswinds, if desired.

LANDING

Once in the air, you will find no difference in the aircraft's handling over a similar tricycle-geared ship. Eventually, however, the landing must be faced, and it comes in two types: three-point landing or "wheel" landing. The three-point or full-stall landing is normally used at first because of its slower touchdown speed and immediate transition to a taxi. The wheel, or tail-high, landing is used for better control in gusty winds because touch-down is made at a higher speed and is quite often preferred by taildragger aficionados.

The three-point landing is achieved by flaring just above the runway and holding the airplane off until the stick is completely back and all flying speed is lost. At this point all three wheels touch down and you are taxiing. If the touchdown is misaligned, expect some nervous moments straightening out the incipient groundloop. If the airplane is allowed to touch down tail-high— a common habit of sloppy nosewheel pilots who haven't done a soft-field landing since their flight test—the tail will drop because of its inertia, increasing the angle of attack and lifting the airplane right back into the air. If you then shove the nose down, contacting the runway at an even higher rate of sink, you'll get another rebound, worse than before. By this time somebody had better get the throttle in hand and initiate a go-around or the result will be one mangled airplane.

When the airplane bounces from a too-hasty touchdown, level off by using a little forward stick, or perhaps by just easing off back pressure, until gravity takes over and the airplane again starts sinking to the runway. If the bounce is quite severe, a little application of power might be wise to keep the airplane out of a stall and soften the sink rate. As the airplane settles back to earth, bring the stick full back to the stop before the tires touch and hold it there during the rollout. Remember, never be too proud to go around—especially in a taildragger.

The wheel landing is a touchdown on the main wheels with flying speed still available, offering better control to battle winds. By keeping the

Executing a tail-high "wheel landing," this Cessna 185 pilot retains good control during the touchdown. As airspeed slows, the tailwheel will sink into the grass, taking over steering duties.

wings at a low or negative angle of attack, no bounce results even though touchdown speed is above normal stall. The tail must be lowered eventually, but you can at least delay this moment of crosswind vulnerability rather than handling touchdown and steering chores all at the same time.

The secret of achieving a tail-high touchdown without bouncing sky-high (the natural tendency, as we've just discussed) is to never let the tail get down in the first place. This means a slightly faster approach speed may be needed, so the flare to stop the aircraft's sink will be achieved without going into a nose-high attitude. With the airplane in level flight just above the runway, only a slight relaxing of back pressure is required to roll the tires smoothly onto the speeding pavement.

There must be little or no sink remaining at the moment of contact, because a sudden vertical deceleration will tend to lower the unsupported tail and trigger a rebound. As speed slows during rollout, more pressure will be needed to hold the tail up, and the tail should be lowered for good steering before rudder control is lost. Naturally, use little or no brakes while the tail is up.

The wheel landing may require some practice, but no taildragger pilot should fly without knowing how to land tail-high. Spring-gear Cessnas, or Pipers with new shock cords, will bounce gloriously if a wheel landing is attempted with an ill-timed shove on the stick in an effort to get the tail back up. Go around or transition into a three-point landing. If you try it again, endeavor not to let the tail get much below level flight before touching down. Don't worry; the airflow will apply more and more pressure to keep the tail down as it rises, so you aren't liable to nose over by wheeling it on as long as you stay off the brakes and avoid soft runways.

So now you're a complete pilot, and you've learned to fly an airplane the way it was a half a century or so ago. Nosewheels are wonderful in stiff crosswinds, and they do make the pilot's job much easier, but the tailwheel airplane will make you a better pilot, and a more confident one. Enjoy it.

CHAPTER
FORTY-THREE

HIGH-PERFORMANCE CHECKOUT

When you're ready to take the inevitable step into larger, heavier, and faster airplanes, it is only logical to get checked out by a qualified CFI. Traditionally, the decision to seek formal training was left to the pilot involved, but some time ago the FAA decided that relying on commonsense checkouts for bigger aircraft types was creating a safety problem too great to be overlooked. Although insurance companies almost universally required checkouts by a CFI, there were still the rare cases of a wealthy lightplane pilot purchasing a P-51 Mustang, crawling into the cockpit, and launching himself into oblivion—all quite legally, of course, because he was operating within the limitations specified on his license—"Airplane, Single-Engine, Land."

THOU SHALT NOT...

FAR Part 61.31(e and f) was added in 1973 to require a logged one-time transition course before a pilot could fly aircraft in two categories: airplanes with more than 200 horsepower and airplanes with retractable landing gear, flaps, and controllable propeller. Although the latter three items are installed collectively in most cases, this brings up an interesting question: Would a retractable-geared airplane with a fixed-pitch prop and no flaps (a KR-1 homebuilt, for example) require a complex airplane endorsement? No, according to FAA sources, because *all three* elements aren't installed.

As regulations go, this one is not particularly burdensome. Any CFI airplane can take care of the requirement, which is simply an endorsement in the pilot's logbook that he or she has been found competent to fly high-performance airplanes. According to current rules, two checkouts are required to meet both halves of the FAR requirement. If a pilot demonstrated proficiency in a retractable with 200 horsepower or less, a 180-horsepower Cessna Cutlass RG, for example, he would not be qualified to

With both retractable landing gear and 250 horsepower, the Trinidad GT requires transition training and logbook endorsements under FAR 61.31(e).

fly a fixed-gear Cessna 182 with 230 horsepower, and an endorsement gained in the Skylane would not suffice for flying the Cutlass RG. However, using a 235-horsepower Skylane RG would satisfy both requirements, as long as two logbook endorsements were made. But gaining a complex airplane signoff in a twin-engine Beech Duchess, with two 180-horsepower engines for a total of 360 horsepower, will not satisfy the over-200-horsepower requirement because of the wording of the rule. Convolutions aside, any sensible pilot would get some qualified person to ride with him when transitioning to a new type of aircraft, particularly if it involves a performance step up, so the regulation hardly takes the place of commonsense practice.

Older pilots enjoy grandfather rights. If they can show they have had logged PIC experience in the over-200-horsepower or retractable-gear airplanes before August 4, 1997, they do not have to seek a logbook endorsement. Also, there is no further requirement for training if the pilot goes into another type of high-performance airplane, such as from a Cessna 182 to a Cherokee Six, or from an Arrow to a Mooney. Rightly so, the FARs assume that a pilot will have received the basic background knowledge for operating a high-performance airplane, and will therefore be able to transition intelligently to other types. It has always been a basic premise of American aviation that judgment cannot be created by regulation, and that the FARs provide only a basic framework to be fleshed out with common sense. For this reason, insurance requirements usually fill the void to protect underwriters against pilots who lack enough judgment to seek a type checkout on their own.

CHECKOUT

What should you expect—or perhaps demand—when seeking this endorsement? Even more so than with primary training, book study must precede actual flight, because you will be dealing with a more complicated machine. Sleep with the aircraft handbook until you know the airplane's systems, performance capabilities, and limitations, as well as the normal and emergency operating procedures. Time spent in ground study will save needless air work, and with the increased expense of flying bigger birds, this only makes economic sense. To make sure you have gleaned the important points from the handbook, a good, stiff oral quiz should be part of the program.

Typical of such an oral was the one covering a Beech Sierra for which a local pilot sought a complex airplane endorsement. I wanted to know the maximum landing gear *retraction* speed, as well as the V_{ge}; although it is not usually an operating hazard for the Sierra pilot, knowing that such a speed exists shows the pilot has done his homework. I also asked how much fuel would be in the tanks when the gauges read "full," another test of his preparation—the Sierra's tanks hold 26 usable gallons apiece, but only 20 gallons are needed to give a full reading on the gauges. To make sure the pilot was familiar with fuel-injected engines, I inquired about the starting procedure for a hot engine. I also wanted to know how the landing gear system operated, and the emergency gear extension procedure. How much cabin payload is permitted with fuel tanks full, and how is it to be distributed? What are the runway requirements under adverse conditions of temperature and elevation? These questions were typical of the oral checkout for a light retractable.

An oral covering an endorsement gained in a heavy-horsepower single, such as a Cherokee Six or a Cessna 206, similarly delves into the systems and operating details. How does loading affect performance? How heavy can the sole two occupants of the airplane be without exceeding forward CG limits? How light can the front seat occupants be before the aft CG limit is reached with a full load of average people? Where is the battery located? How does one manage the fuel tanks of the Cherokee Six? *Know thy airplane* is the criterion, and more airplane means more to know.

Because a more complex avionics package is normally part of the higher-performance airplane, some time is spent on radios; puzzling over the audio panel in a faster airplane can get a tyro pilot into trouble before he realizes it.

To become familiar with one new power control, the constant speed prop must be seen as a "gearshift" control; low gear, or low pitch, provides high rpm for takeoff, a second "gear" selects a cruising-climb setting, and a high "gear" or higher-pitch setting is used for low-rpm cruising power. The power lever—which is still the throttle—is moved to a lower-power setting before each reduction of rpm so the slower-turning engine will not be "lugging" with nearly full throttle. Manifold pressure is the only reference available for adjusting the throttle because the propeller governor will maintain a constant rpm as power is changed. The propeller control is returned to low pitch (high rpm) position just prior to landing so that the engine will be ready to respond to full throttle in the event of a go-around.

During the actual in-flight training I watch to see if the transitioning pilot is learning to keep up with the faster plane. Bigger airplanes with heavier wing loadings tend to be more stable than light trainers. This makes them somewhat easier to fly, but their very stability can cause them to sink right on into the ground if the pilot gets behind the airplane on a landing approach.

Flight maneuvers used to transition the pilot to high-performance flying include such old standbys as steep 720° turns and slow flight. The steep turns increase the critical nature of control inputs to speed up the process of getting the feel of the airplane, whereas slow flight teaches the pilot what to expect in the traffic pattern, particularly when combined with flap and/or landing gear extension while holding airspeed and altitude.

Stalls are a natural outgrowth of slow-flight practice, and must be demonstrated from all normal flight conditions. Usually the meanest configuration will be a partial-power, full-dirty landing approach stall, a situation where a high-performance airplane usually gives little warning of imminent disaster until the bottom falls out. Instrument flight is also worked into the transition, commensurate with the pilot's ratings; a VFR private pilot will need to demonstrate only basic survival skills; an instrument-rated individual will get some approach work.

AROUND THE PATTERN

Takeoff and landing practice is the payoff, and nowhere is a tendency to get behind the bigger, faster airplane more evident than in the traffic pattern. Increased torque makes right rudder application more pronounced during takeoff; failing to anticipate this causes frequent deviations from the centerline. A climb-power reduction takes a extra second or two because there are now two controls to be adjusted, and a fast climb rate brings the airplane up to altitude quickly—often before the new-to-type pilot is ready for it. Upon reaching pattern altitude, power should be reduced for the downwind leg so the airspeed will be kept compatible with normal training-aircraft speeds to avoid conflicts in the pattern and make it easier to transition to the landing approach.

The downwind leg, or the initial pattern entry just prior to descent if a base leg or straight-in approach is made, is the place to have a habitual prelanding routine established. GUMP is still a good short checklist: G—gas on proper tank and boost pump on if needed; U—undercarriage (gear) knob moved to the down position; M—mixture control returned to full rich if not already done; and P—propeller pitch set to high rpm (some pilots prefer to leave the prop alone until short final to avoid an increase in noise level).

After running the GUMP check, a power reduction is made to an appropriate approach setting, usually 10 to 12 inches of manifold pressure, to allow a normal-size pattern to be flown. Power approaches are *de regle* because power-off glides would be short and steep in these highly wing-loaded airplanes. At the time of the first power reduction a check of the gear-down lights should be made to see if the gear has completed its cycle. Speed is reduced to a figure equivalent to 1.3 times the stall speed in landing configuration, plus 10 knots for maneuvering. Partial flaps are extended no later than the base leg, or about 2 miles out on a straight-in approach.

On final approach, speed is reduced by adding full flap, and a stabilized approach is sought; speed should be constant at 1.3 V_{so} with power adequate to carry the airplane to the runway without further adjustment. One should strive to avoid letting high (over 500 fpm) sink rates develop on short final, a sneaky condition that can cause a high-performance airplane to sink through the bottom of a last-minute flare to make a hard touchdown.

A few hours spent in initial transition training for high-performance airplanes is well worth the time, because the pilot who has not learned to stay ahead of his airplane is an accident going somewhere to happen. Take advantage of such a program to sharpen basic flying skills, and use the same checkout procedure whenever moving into another complex airplane, even though further checkouts might not be required by the FARs.

FLYING FOR MONEY

Most private and student pilots know the next level of pilot certificate beyond the basic private license is entitled "Commercial Pilot." The commercial certificate is required if pilots are to receive money for their services, but the rating also is a laudable goal for anyone interested in bettering her pilot skills, whether or not she might actually be contemplating a career in aviation.

Requirements for the commercial certificate are outlined in FAR Part 61.123. As you will discover, you must be at least 18 years of age, have earned a passing grade on the commercial knowledge exam, and have logged 250 hours of various types of pilot experience. You do not have to obtain the second class medical exam unless you intend to actually fly for hire. I always encourage prospective commercial students to take the knowledge exam first; having the written passed places you under a certain time constraint because it expires in 2 years, which encourages aggressive flight training.

It was not always necessary to have 250 hours for the commercial ticket; as a matter of fact, I took my first charter trip at the tender age of 222 hours. The minimum time was increased from 200 to 250 hours in 1973, when FAR Part 61 was extensively revised. This was done to assure that commercial airplane pilots would add enough hours to get an instrument rating, which became mandatory for unrestricted commercial flying at that time.

It is not true, however, that an instrument rating is an *absolute* prerequisite for a commercial ticket. Recognizing that some paying jobs can be done safely without instrument training—such as aerial application, sightseeing rides, and local photography flights—the FAA retained a provision for a limited VFR commercial license that allows carriage of passengers for hire within a 50-mile radius and during daylight hours only.

For unlimited commercial privileges, however, an instrument rating is still necessary, and most professional flying jobs are closed to noninstrument pilots. It is wise to work the instrument curriculum into the program first, particularly if you haven't quite accumulated the 250 hours needed for the commercial; then go to work on the commercial pilot certification. On the other hand, if you are seeking a commercial ticket for personal purposes only, you might wish to save more than half the cost by going after a limited commercial.

COMMERCIAL DEFINITION

When do you actually need a commercial certificate? Basically, if you get paid for flying the airplane, you are operating commercially and need the license. If you are a traveling salesperson, for instance, and the company doesn't *specifically* tell you to cover your territory in an airplane, you may legally make your rounds on a private ticket. Likewise, you can demonstrate an airplane to a prospective purchaser on a private license, because then you are a salesperson and not a hired pilot. However, if you are required to haul cargo or personnel for your company, even though no charges or bonus remuneration is involved, you are being paid to fly and should have a commercial ticket.

Don't assume the acquisition of a commercial certificate gives you the right to haul people around the country for hire. You will also need an FAR Part 135 air taxi certificate, which means your entire operation—including airplanes, flight manuals, offices, and record keeping—must be inspected periodically in addition to being staffed by qualified pilots. Remember, the operator must hold an air taxi certificate while you as the pilot must hold a commercial certificate; both requirements must be met to fly a charter. As a crowning blow, you cannot fly commercially unless you are covered by an FAA-approved drug-testing plan. When Congress mandated the FAA to

Flying this corporate Turbo Commander twin is a fine job for a commercial-rated pilot. However, there is also satisfaction is holding the commercial license even if you never fly for a living.

assure that commercial flying was drug-free, the agency took it upon itself to require drug monitoring for *any* carriage of paying passengers, even a few sightseeing rides.

If you are an airplane owner, you may receive a break on your insurance premium with a commercial ticket, although it probably won't be enough to pay for the training. Mostly, you'll take pride in the fact that you've chosen to upgrade yourself to a higher standard of demonstrated proficiency. In truth, a very sharp private pilot with 50 hours could probably perform the commercial maneuvers well enough to pass the checkride, but there's more to flying commercially than just getting twice as good as you were for the private ticket.

REQUIREMENTS

What else is needed? Disregarding the instrument rating, the underlying commercial requirements are for at least 100 hours of pilot-in-command time, 50 hours of which must be cross-country time, which means you landed at least 50 nautical miles from the departure point, including one solo 300-mile trip with landings at three points, one of which is at least 250 nautical miles from the departure point. At least 20 hours of dual instruction are needed, with at least 10 hours of it as instrument dual and 10 hours in a complex airplane. Both day and night dual cross-countries must be logged and at least 3 hours of recent preparation are needed for the commercial flight test. At least 5 hours of solo night flying time are needed, including 10 takeoffs and landings under tower control.

Aside from the instrument rating curriculum, which you may complete earlier or skip entirely, it would be normal to expect to run up 25 to 30 hours in preparation for the commercial flight test. The concurrent ground training will concentrate on detailed knowledge of aircraft performance, systems, and loading requirements. During the oral portion of the exam, expect to be grilled thoroughly on every aspect of safe piloting; if you're going to charge for your services, you are expected to be worth the pay.

ADDITIONAL SKILLS

In addition to most of the tasks performed for the private certificate, done to a slightly higher standard, there are a few new maneuvers on the commercial test. These consist of a power-off accuracy landing, a steep 1,080° spiral, an eights-on-pylons maneuver, and the time-honored lazy eights and chandelles. The spot landing is deceptively simple; close the throttle and maneuver to land no more than 200 feet beyond a specified point.

The steep spiral involves flying a set of descending constant-radius turns around a point on the ground, recovering to a heading to simulate a traffic pattern entry. Eights on pylons are a specialized maneuver in which the wing tip is kept in constant alignment with the pylon during the turn by holding the airplane at a pivotal altitude determined by ground speed. Lazy eights are a test of one's ability to fly a predetermined pattern of 180° turns above and below the horizon in reference to a line from a fixed point, using smooth and delicate control inputs. The chandelle is a steep 180° climbing turn, using the stored energy of the airplane's speed to gain the maximum altitude possible as airspeed bleeds off during the turn; this is an exercise in planning, demonstrating your ability to think ahead of your airplane in order to make it arrive in a certain flight condition at the end of the maneuver.

The maximum performance approaches and landing on the commercial checkride also demonstrate planning as you attempt to place the aircraft in a normal touchdown beyond and within 100 feet of a given line. Speed control, drag adjustments, power reduction, and turn placement all must be balanced carefully to bring the aircraft down on target.

The old standbys—cross-country flying, crosswind takeoffs and landings, soft and short-field operation, slow flight, steep turns, stalls, and forced landings—are all in the commercial curriculum, with slight changes added to test increased skill and with performance tolerances more or less halved as compared with the private checkride. Expect distractions from the examiner, who will attempt to place you under the stresses paying passengers might generate. It will also be necessary to demonstrate proficiency in a complex airplane (retractable gear, flaps, and controllable propeller) on the flight test, normally just a few minutes of basic and emergency maneuvers, although there is nothing to prohibit taking the entire checkride in the complex airplane.

For the dedicated and serious student of aviation, the commercial license is not too difficult to attain. However, it takes a certain amount of desire to improve yourself, and although you may not have visions of flying for a living, it's nice to be able to hold in your hands a tangible reward for your efforts. Well-paying jobs in aviation are scarce and highly sought after, but the ability to fly for your company might make extra points on a résumé, even in nonaviation employment. Consider adding the next level of competence to your license, especially if you're already instrument rated. It's a great exercise for jaded flying skills.

INSTRUMENT RATING— WORTHWHILE?

Eventually, most private pilots who are reasonably unimpeded by financial constraints will consider getting an instrument rating added to their licenses. After being weathered out of a few choice trips while other pilots depart through a cloud layer, or hearing a fellow pilot brag about the joys of IFR flight, it's only natural to want to join the club.

Before you plunge off into pursuit of the instrument ticket, however, it would be a good idea to analyze the cost/benefit ratio in your own operation. Many new instrument pilots are disappointed with the increase in utility they have gained, and they wind up flying very few IFR hours despite having the legal ability to do so. For some pilots an instrument ticket will be an under-utilized luxury, whereas for others it is a virtual necessity. Let's see which category comes closest to matching your situation.

Certainly an instrument rating would be a worthwhile goal if you have time and money to spare for self-improvement. You will gain valuable insight into controlled-airspace operation, which will make you a safer and more poised pilot. Your weather wisdom will be increased and you will possess a valuable "escape" tool in case you get suckered into a bad weather situation. If you plan to go on to seek a Commercial Certificate someday, you might as well plan to get the IFR ticket because an instrument rating is nearly a prerequisite for a working pilot. To qualify for the instrument rating, it is necessary to have accumulated 50 hours of cross-country flight time as pilot-in-command, so it pays to make every flight include a landing more than 50 nautical miles from the departure point until you've logged the requisite hours.

RATING REQUIREMENTS

These advantages notwithstanding, the fact remains that a lot of instrument pilots don't use their rating enough to justify its cost. You should figure on spending at least as much money for instrument training as you did to get the private ticket, if not a bit more; a more expensive airplane (probably) and avionics package (absolutely) are used for instrument work. **A total of 40 hours of instrument time is needed to qualify for the rating,** at least 15 hours of which must be dual, although it has been my experience that somewhat more than 25 hours of dual are required to make a capable instrument pilot. Half the 40 hours can be logged as dual instruction at the controls of an approved ground simulator, saving valuable airplane time while you develop a scan pattern and learn how to fly an approach for the first time. However, there's no substitute for the *real* airplane, in *real* weather.

Instruction must be given and proficiency demonstrated in at least three types of commonly used instrument approaches; in addition to a precision approach like the ILS, two nonprecision types must be flown, such as RNAV, VOR, and ADF. Therefore, the airplane must have enough equipment to fly these two procedures (or a suitable simulation must be provided). At least one cross-country flight covering more than 250 nautical miles must be included in the training, utilizing three types of approaches, each at a different airport. You should plan on flying between 2 and 3 hours at a session when you're immersed in instrument training to permit more approaches to be practiced and to allow for shuttling between airports.

First, however, there is a knowledge test to be passed, something that might as well be tackled before flight training begins in earnest. There's nothing tricky about the test, but you do have to know the material. You can prepare equally well with a home-study course, an integrated FBO

This cloud deck is almost solid, but the pilot with IFR capability can fly on without worrying, knowing that the weather ahead is within his or her limitations.

program, or computer-based preparation materials, whichever suits your study habits. You might as well buy at least a one-time set of NOS charts and approach plates or other commercial equivalents unless you're ready to spring for a subscription to Jeppesen coverage that the Big Boys use. With the cost of training, equipment, and charts, the price begins to add up.

WHAT THE RATING WILL AND WILL NOT DO

Now, consider what you will be getting for your investment. Too many VFR pilots think their lives will become simpler with an instrument rating, because they can fly with fewer restrictions. *Wrong*. Life gets *more* complicated, not less so, because more and weightier decisions must be made. In return for the privilege of flying through clouds, you sacrifice your freedom to depart, terminate, and alter your flight as you choose. Each action is carefully controlled to preclude conflicts with unseen traffic. You *ask* for a clearance to depart, for a change in altitude, for a new routing; less spontaneity is possible.

Likewise, it is all too often presumed that weather can no longer delay a flight. *Wrong again.* The instrument aviator can fly in lower weather minimums, but there are still weather minimums, and there is still plenty of weather bad enough to cause a cancellation; zero-zero fog halts even airline traffic. Even if the weather is up to the 200 feet and one-half mile of ILS minimums, when you're a new or rusty instrument pilot you may want to wait for 500 feet of ceiling while you're gaining your precision. Thunderstorms, icing, low-level turbulence, widespread near-minimums weather with no alternate airports in reach, or an airplane with limited avionics—these are just a few of the things that can keep an instrument-rated pilot on the ground despite the rating in his pocket.

Remember, little airplanes are really meant to fly only in warm stratus clouds. Turbine-powered equipment, on the other hand, can expand a pilot's range of altitudes and offer a fast climb rate through the lower crud, advantages we little-airplane people don't have. The airlines fly in almost any weather conditions, but remember that they use *two* pilots, with the best available equipment, over familiar routes, with extensive ground support— and they *still* overfly or cancel some stops at times. You, in a single-pilot, limited-performance, random-routing operation, must also expect to let good judgment cause some cancellations, even with an IFR ticket.

WILL IT PAY?

If you fly quite a few cross-country hours in the course of a year, or live in a part of the country subject to frequent stretches of fairly localized below-VFR weather, you'll certainly benefit from the instrument rating. You can fly perhaps 75 percent of your trips without it, and maybe 90 percent with it; to achieve the other 10 percent will require known-icing certification and turbocharging.

Add up the cost of getting the rating; include the time lost from work, the actual training expense, and perhaps the cost of ferry flights to and from a suitable instructor. To this add the tab for any additional equipment and inspections required to bring your airplane up to IFR standards, if the airplane isn't already in shape. Don't forget the expense of instrument charts and the periodic refresher training needed to stay current. From the total thus far you can deduct any insurance premium savings resulting from adding an instrument rating.

Now estimate the number of extra trips you *honestly* could have made last year, given the capability provided by your airplane's performance. The importance of these additional flights determines if it is practical for you to go after an instrument ticket. If you're short of loose cash to spend in getting your airplane up to speed, or if you don't plan to fly all that many hours cross-country, maybe you'll opt to stay VFR for now. You need not apologize for your decision, and you can take comfort in your simpler lifestyle.

On the other hand, the possession of an instrument ticket does give you an extra option to play. You can press on with deteriorating weather conditions just a bit longer than a VFR pilot, knowing that you can pick up the microphone and get a clearance. You can leave a socked-in airport a few hours earlier when conditions first start to improve, rather than waiting for the sun to break out. Perhaps best of all, it is a considerable comfort to know that you can always leave this forlorn weathered-in field if you want to work at it hard enough; the VFR pilot has no such ability, and whether or not you always choose to exercise the capability, it's nice to have it.

The additional knowledge and experience of instrument training is worth a lot. If you've never seen the world change from dingy gray to bright and sunny with a few thousand feet of altitude, or had the pleasure of being in perfect alignment with the approach lights as you slide out of the overcast, or enjoyed the luxury of spending a damp, rainy night at home with the family rather than somewhere in a motel, you can't really appreciate all the benefits of flying by instruments. Just remember the price tag, and decide if it's for you.

IFR ITINERARY

Congratulations—you've struggled through the knowledge exam, a long, tiring flight training program, and a white-knuckles checkride. Now you possess an instrument rating and you would like to put it to work, but you don't quite know how to begin. You have been given the right to shoot a 200-and-one-half approach, but not the confidence to tackle it alone.

Well, join the club. Every new instrument pilot has had to go through the same things. Flight schools are very good at pumping out quick instrument ratings, but they often graduate students with no actual experience in flying through weather. Those who need an instrument ticket are usually busy, involved people, and they must take a concentrated course of instruction in order to complete the training at all. Unfortunately, this can lead to minimal competence and even less confidence, because the statistical probabilities are against encountering actual instrument weather in a short training program. There are no substitutes for hard rain on the windshield, a dark, wet cloud, and the trapped feeling of nearing minimums while still shrouded in stratus. It's different than when wearing a hood in the sunshine with a professional in the right seat, no doubt about it.

And yet you got the instrument rating to use, not to sit on, so you must get to work immediately gaining solo proficiency. Using instrument flying skills aids retention of the hard-won ability and builds confidence. Putting off that first solo venture into a cloud only makes it harder to face the inevitable, so don't wait even a week to start filing IFR, if possible. Naturally, there's a proper way to begin—by easing into it slowly.

FIRST WALK, THEN RUN

When tackling your first local instrument approaches to build proficiency, establish minimums somewhat higher than those allowed by the approach plates; for nonprecision approaches you might wait for conditions of 1,000-foot ceiling and 2 miles visibility, and even for a full-ILS arrival you should have 500 feet and 1 mile for openers. This is not because you can't fly well enough to keep the needles centered if all goes well, but to cover yourself if you get slightly behind the airplane while handling the communications chores. In such a case, you could be forced to miss the approach unless you have marginal VFR underneath. Don't stick to clear days, however; you won't build confidence by filing IFR in the clear.

So with your pencil poised and your charts neatly folded, call for that first clearance solo; you'll probably find it easier than you thought. Unless you're operating in a terminal area, there will be few other airplanes on the frequency, and you'll receive immediate response to your call, unlike the hectic weekend traffic on VFR channels. Don't forget to turn the transponder on before takeoff, have the radials and frequencies pretuned before entering the clouds, and enjoy yourself. Breaking out into the clear and looking back at what you've come through is what it's all about, and it's a great feeling.

DEPARTING CROSS-COUNTRY

After polishing your cloud-flying skills close to home, prepare for IFR cross-country work. Hopefully you will have been making all VFR cross-countries on an IFR clearance to gain familiarity with ATC procedures as pilot-in-command. When you're ready for the real thing, don't bite off more than you can safely chew the first time out. A 3-hour cross-country in the clouds with no autopilot is tiring even for an experienced pilot, especially when mixed with turbulence and a gnawing doubt about the wisdom of being there.

For your first instrument exposure beyond the local area, wait for a day when your departure airport is just beginning to open up and good weather is waiting a half hour away. Without an instrument rating you would be forced to wait until conditions improved, but with the rating you may as well depart as soon as you have enough ceiling to organize your act before entering the clouds. As you are no doubt aware, noncommercial flights can legally depart under zero-zero conditions, but doing so is a calculated risk for competent pilots, because of the inability to return if something goes wrong. Best to wait for a 500-foot ceiling and establish a definite improving trend to back up the reports of clearing conditions en route. Knowing it'll get better as you go along always makes the trip seem shorter.

Do not be satisfied with a few minutes of actual IFR while climbing to VFR conditions on top of the clouds. If you can see that a lower altitude will put you back into the stratus, ask for it; you need the practice, and you can always go back up into the clear if you feel the need to take a break. Don't plan on the tops being right where they were, however; the effects of terrain and air mass mixing can lift the cloud tops as the flight progresses. The tops are always a thousand feet higher than expected, and the icing level a thousand feet lower.

As your proficiency and confidence increase, you should begin filing into actual instrument conditions from a VFR departure, shooting

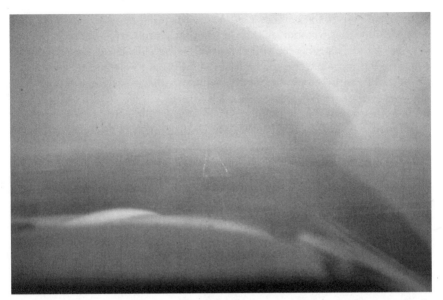

There is no more welcome sight than that of a runway appearing out of the clouds at the end of an ILS approach. Work up to full proficiency in gradual steps.

approaches from an en route environment, rather than the local approaches you have been executing to maintain currency. Always plan on having sufficient fuel to return to good weather, even when an alternate isn't required, just in case the destination weather goes down or an equipment failure prevents you from making the approach. In fact, you should have *two* alternates in mind, the one designated in your flight plan and another to back it up. Never place yourself in a position with no out. Keep track of conditions at your alternates as well as the destination and you'll always have landing minimums somewhere within reach.

Finally, after first filing from bad weather into good, and then from good conditions into bad, you can start making intentional flights involving IMC from beginning to end. Keep your minimums-for-dispatch weather reasonable; don't blast off just as soon as the conditions along your route rise to landing minimums. Wait for a trend to be established that will give you reasonable safety and assurance that you can conduct the flight as planned. If you try to depart as soon as you get reports of bare-minimums weather, you run a good chance of having one or more of your key airports slip back down, leaving you with disrupted plans or perhaps no alternate within reach.

WATCH THE WEATHER

Learn to maintain an en route weather watch. An instrument pilot has an excellent opportunity to watch weather develop, especially if she is on top or between layers; she can correlate this with the reports received. By continually checking weather at the destination and alternates, and applying your firsthand impressions, you can avoid being placed in a low-fuel situation or continuing into unsafe weather. A VFR pilot may be able to proceed on eyeball weather only, but an instrument pilot must supplement what she sees with fresh METARs, amended forecasts, and current pilot reports if she wants to avoid nasty surprises.

There will always be weather that must not be tackled, even if you are current and confident. Thunderstorms are bad news and are usually best handled by staying VFR under the cloud base, where you can see and avoid the rain shafts. Don't make a habit of relying on ATC radar to guide you around embedded cells; someday the service might not be available. Likewise, the hazard of ice is better avoided than battled. Stay below the freezing level in clouds, climb into the clear air on top if able, or return to warm air behind you. The most important rule to remember when encountering icing conditions is *do something; don't just sit there.* Change your route or altitude to fly where the ice isn't, or don't fly at all. The final bane of instrument flying, dense ground fog, can be easily avoided by always having enough fuel to reach good weather—and knowing where the good weather is. Land short and top off the tanks if it appears fog might be a factor at your destination. Knowing you can cope with these weather hazards by making the right decisions increases your confidence in your ability.

To grow and improve as a green instrument pilot takes practice, not just simulator time and hood work. You must get your wings wet in the actual ATC environment and continue to do so as long as you seriously want to keep your instrument proficiency. So if you're new to instrument flying, or if it has been a long time since you flew in IMC, take an experienced hand along the first time to help handle the right seat chores, but get that hands-on practice you need.

FILE IFR OR STAY VFR?

I was being chided by an experienced fellow pilot because I had arrived at a meeting by flying VFR under a low overcast. "LeRoy, you ought to know better than to go scud-running like that!" he admonished. I knew that whenever he encountered weather on a cross-country trip, he preferred to bore on through rather than avoid it. To him, the decision was always clear: File IFR or don't fly at all.

I accepted his advice in the spirit it had been given, but thought to myself how quickly we often judge someone's actions without considering the circumstances. What appears to be a dangerous, thoughtless action might be a reasonably safe and correct choice for a particular pilot, in a particular plane, on a particular day. The limitations of both the pilot and the machine must always be considered when deciding the best course of action.

HOW TO HANDLE WEATHER

Thus it is with the question of how to handle weather. In training we tend to define weather as either good-VFR or hard-IFR, forgetting that in the real world we seldom enjoy such clear-cut definitions of weather. Most of the time inclement weather can be flown as either marginal-VFR or easy-IFR—such conditions as 1,500-foot ceilings with 5 miles visibility, or 1,000-feet and 10 miles, or an unlimited ceiling with 3 miles visibility in haze. In these borderline cases, the pilot must make a decision about how best to handle the weather: Should he attempt VFR flight, or file IFR if he is instrument rated?

We cannot overemphasize the need to *make* a decision. Murky weather is no time for vacillation; you need to make up your mind what you are going to do before you take off, and update your plan as the flight progresses. Sitting there scratching your head as you dodge the barely VFR soup is silly, as is canceling your IFR prematurely and later wishing you hadn't. To make intelligent decisions, you need hard information about the weather situation, so get a good briefing before takeoff. You need to know where the weather is worsening and where it is improving, what sort of system is affecting your route, and what is expected to occur with the passage of time. With this briefing, you can apply your observations and the METARs to make assessments of your ability to complete the flight.

Go in the proper frame of mind to tackle weather. Forget about getting somewhere on schedule, or even getting somewhere in particular at all. Decide only if it is practical for you to set out along the route and be ready to divert without hesitation if the weather seems to be running behind the forecast. You can always reshuffle your business appointments or vacations— as long as you're around to do it.

ARE YOU IFR OR VFR?

If you are not instrument rated or your airplane isn't suitably equipped, your life is less complicated. You simply should not begin any trip unless there is a reasonable assurance that you can get there free of clouds. Beyond this basic need, you must add enough extra margin of cloud clearance and visibility to accommodate your navigation ability, your familiarity with the route, and the time of day.

If you are instrument rated and equipped, you have the option of penetrating the weather if you so choose. Although IFR may seem to be the way to go whenever conditions slip from CAVU, there could be reasons to avoid the ATC system. Not every airplane is able to handle every type of weather; a high-performance airplane with a turbocharger and an oxygen bottle allows one to tackle conditions that would be hazardous in a light single. Perhaps the airplane you're flying has just begun to exhibit radio problems; it would be poor judgment to file into weather without being sure of your basic avionics. Perhaps there are reports of icing at the levels you would be forced to fly, and you are not equipped or inclined to tackle ice. Or radar may be reporting thunderstorms mixed into the soup. If you've recently had a brush with a cumulonimbus you may be unwilling to file IFR, preferring to avoid the storms visually.

Looking over the weather situation carefully allows the IFR pilot to make a decision about the mode of departure. If marginal VFR conditions exist locally, you may wish to depart IFR and seek a comfortable flight level on top or between layers, with the option to descend to VFR conditions if you encounter hazardous weather aloft. If you suspect dangerous weather at the altitudes required for IFR flight, you may want to leave VFR and go underneath, with the option to pick up a clearance if you encounter lower conditions en route.

If you plan on climbing up through a broken layer to VFR-on-top, you may as well pick up a clearance to your destination; things have a way of getting busy if you just cruise into a terminal area VFR-on-top, and entering the system from a distance will smooth the handling of your arrival.

And, if the weather situation worsens ahead, you're already in contact with ATC to expedite your options.

SNEAKING OUT? BE CAREFUL

If you plan to leave VFR below the overcast, take care to choose a route that will permit you to plug into the ATC system later if needed. Sneaking up the valley VFR might not be so smart if you need to talk to Center in a hurry and you find you can't get high enough to make yourself heard. Or a direct course may take you so far from a communications outlet that voice and radar contact is impossible. If you remotely suspect you will need to file IFR, stay close to a route with ATC communications coverage, or airports with telephone service.

Next, don't wait too long to file. We're getting all too many reports of instrument-rated pilots involved in marginal-VFR weather accidents, evidently because they thought they could make it on through without filing. Just as you did in your VFR days, tell yourself how low you're letting yourself go before you chicken out, and *do it.* The minimums are up to you; if you aren't familiar with the country where you are, you'll probably give up sooner than you would if you knew the landmarks or were only a few miles from home base. Just eyeball the elevation of the terrain and obstructions, assess your VFR skills, and say, "When I'm getting pushed down to *this* altitude, I'm getting out of here."

No two pilots will probably handle the situation in exactly the same manner. One of the saddest predicaments is an airplane with two captains, both of whom are equally experienced with nobody really in command. Often, the pilot in the left seat will head off into weather he really doesn't want to fly because he doesn't want to lose face in front of the other pilot. If I'm riding the right seat, I make sure the pilot knows he's in command and I'll ride with him, abiding by his decisions. He can ask my advice, but I'm not going to fly the airplane for him, and I don't want him to worry about what I'm thinking.

WHEN NOT TO FILE

The heated discussion at the beginning of this chapter involved a stalled-out cold front over relatively flat terrain. I was flying a well-worn Skyhawk with vintage radios, equipped for localizer-only approaches, and the surface temperatures were only 3 or 4°F above freezing. Tops were expected to be around 9,000 feet AGL. I was certainly not prepared for the ice I would get trying to climb above the weather, and the 'Hawk would run out of climb ability just as I was getting into the top of the cumulus where the ice is

With ice in the clouds overhead and limited climb capability, this trip down the highway had better be made underneath the weather, even though the visibility is just over 3 miles.

often heaviest. Compounding the problem, there was no open weather nearby, only a promise of improvement at the destination.

The route under the clouds, by comparison, was well populated with VFR airports where I could seek refuge, and an FSS outlet was available for consultation in the middle of the route. So I elected to make this trip *under* the weather rather than through it. I did have a minimum altitude chosen, about 700 feet above the highest terrain elevation in this case, because visibility was relatively good below the cloud bases. At one point I was pushed almost down to my minimum and I was preparing to execute plan B—a 180° turn to land and file low-level IFR—but the clouds soon rose as reported and it was not necessary.

For me, it had been a safe and easy trip. I was comfortable with the route and weather, but not everybody would have been, including my IFR-only friend. His pilotage navigation skills would have been rusty, and he would have been much more comfortable in the clouds with his radios and autopilot. Both of us were right, each in our own way. The best advice I can offer is to never continue with a flight if you don't feel comfortable doing so. Respect your own limitations—don't use someone else's.

C H A P T E R
F O R T Y - E I G H T

REGAINING IFR CURRENCY

If you are an instrument-rated pilot, you are no doubt aware of the requirement for recent IFR experience, as defined in FAR 61.57(c). Unlike some other recent-experience requirements, this FAR does not apply just to passenger-carrying privileges; you cannot fly under IFR or in the clouds, solo or otherwise, unless you meet the currency standards of 61.57(c). You must have logged six instrument approaches, actual or under the hood, within the past 6 months, along with some holding, intercepting, and tracking.

An approved flight simulator or flight training device can be used to acquire this time, or you can log the experience while flying in actual instrument weather conditions as long as you're current. If the in-flight time was logged using a hood or other view-limiting device, you must show the name of the safety pilot in your logbook, although he or she need not personally sign the book. This safety pilot must be seated at the controls, hold at least a private pilot's certificate for the aircraft exam, and have a current medical exam.

If you've neglected to put in the specified time during the past 6 months, there is no requirement that any dual be sought to regain instrument currency, as long as no more than six additional months have elapsed since you were last current. You can take your hood and safety pilot, fly off six instrument approaches and perform your holds, and be once again legal for IFR. You might not be safe, but at least you are covered under the FARs. However, following two 6-month periods, during one of which you were still current and during the other eligible for self-renewal of currency, you are barred from operating under IFR until you take an instrument proficiency ride with an instrument instructor.

COMING BACK

So you look in the old logbook one day and see that you have had good luck during the past 6 months, avoiding IFR weather to the extent that

259

you have not accumulated the required six approaches. Because you were current until recently, you have the option of getting the practice in on your own, without an instructor. Many instrument pilots prefer to take a period of dual and get an instrument proficiency endorsement instead of doing it themselves, particularly if it can be combined with a flight review. However, there is nothing wrong with self-improvement, properly done.

To accomplish this, a pilot needs an instrument-equipped airplane, a VFR day, a safety pilot, and a hood. The airplane need not actually be IFR certified, as long as the equipment is in good working order, because you will not be legal to enter instrument meteorological conditions anyway, unless the safety pilot is qualified and willing to assume command. Bear in mind that transponders must be certified every 2 years whether used for IFR or VFR, whereas the altimeter and static system check and the VOR receiver test are needed for IFR purposes only.

Choose your safety pilot wisely. Would you trust your life to this person? That's exactly what you might be doing in relying on him or her to keep you separated from other traffic and deteriorating weather. If possible, a knowledgeable instrument pilot would be a wise choice for a safety pilot, someone with whom you can share recurrency training duties. He or she may be better at keeping abreast of the situation and anticipating your moves, whereas a VFR pilot would know very little about your intentions. If you think this person could use a backup, put another pair of eyes in the rear seat to help spot traffic.

Stuck in this below-VFR weather, this pilot vows to work on regaining his ability to fly instruments before his next trip.

WHO DOES WHAT

Cockpit duties should be carefully defined before flying; a too-helpful safety pilot is not useful for regaining proficiency, but she should be made aware of what is expected of her. You should assume all communications chores relative to the IFR flight unless you are losing control of the airplane and the controller needs an immediate answer for safety reasons. Only then should your safety pilot help out. However, VFR traffic advisories are the safety pilot's responsibility, and she should be ready to handle the "Got 'im" reports. She should not bother you with "We've got one off to the right about a mile," unless avoidance maneuvering is likely; if it's just a topic of conversation, you've got enough to do under the hood without the small talk. Naturally, if she says "I've got it" you will release the controls to her, but wait for an acknowledging shake of the control column so you know she's talking about the airplane, not the traffic she's just spotted.

As long as the aircraft is legal and your safety pilot is a current IFR aviator, you can operate on an IFR flight plan with her as PIC; you no doubt need practice in operating within the system, not just keeping the airplane right side up. In addition, more expeditious handling will be given your requests if you are operating on a clearance, rather than popping up VFR to shoot an approach.

Plan on getting in as many varied types of approaches as possible: ILS front course, ILS back course, GPS/RNAV, VOR circling, VOR straight-in, VOR/DME (distance measuring equipment) arc, NDB (nondirectional beacon)—the more the better. It generally takes more than one approach to get one's confidence renewed, so if the traffic is light you can try to get approval to shoot the ILS to minimums, pull up straight ahead to fly out the back course, make the procedure turn to do the back course approach, then proceed out the front course for another trip down the glideslope. If no back course approach exists you'll have to be satisfied with vectors back to the ILS final approach course. Repeat as needed, but don't quit on ILS approaches. Go over to a smaller field with a GPS or a VOR circling approach, and put some precision into your nonprecision approaches.

NO-RADAR PROCEDURES

One by-product of these boondocks approaches is that you might get some experience in a nonradar environment, rather a shock after being nestled comfortably in the lap of center radar. "Radar service terminated" comes as a rude awakening. A clearance to hold and "Expect further clearance at…" means another airplane is out there somewhere. With poor radar and communications coverage, arrangements have to be made for recontacting ATC after the missed approach. It's all good review for the rusty aviator sweating under his hood.

After making an abeam passage of the VOR outbound, despite one's best efforts, one resolves to do better on the inbound leg. The procedure turn can be interesting with a bit of wind, and this knowledge had best be incorporated into the planning for timing the approach, anticipating a tail-wind or headwind while coming down to the minimum descent altitude (MDA). If, like many pilots, you steam along outbound at full cruise, then slow to an initial approach speed during the procedure turn, you will wait and wonder just when that approach fix is going to come up. Finally, it comes; you bomb down to MDA quickly in order to break out before reaching the 1-mile circling radius, and struggle to hold MDA precisely while the time runs out. Don't forget to sing out on the CTAF and keep your safety pilot awake. Don't bust through a traffic pattern needlessly; pull up for the miss before penetrating the touch-and-go pattern if at all possible.

NDB approaches are equally fun, sometimes bordering on the mirthful before you begin to wonder what you would have done if the weather had been IMC and you missed the final approach course by 2 or 3 miles as you did today. Pin down a wind correction angle inbound on the initial approach and outbound for the procedure turn, then try to adjust your turn legs to bring you out onto the final approach course with the proper relative bearing to hold the course. Lotsa luck, and remember that this could be the only approach available someday.

THE FAILURE MODE

Don't use your autopilot excessively; let it hold the heading and altitude while you unfold charts or brief the approach, then hand-fly the rest of the way in. It's important to practice its operation, but you should never know-ingly place yourself in a position of needing the autopilot to get through a situation you couldn't handle on your own. It's a wise practice to simulate some equipment failures now and then, just to see what disruption it might cause in your routine. You never know how much you'll miss that second nav head until it goes bananas in the clouds. Partial panel work also deserves attention, against the day a gyro or vacuum pump goes to sleep.

Bringing yourself back to full currency is not just a matter of flying straight and level down the airway with the hood on; it's giving yourself a real-istic workout. Plan on flying approximately 2 hours of approaches, as well as a little en route work. By doing so, you will probably discover some shortcom-ings, and learn that currency is a relative thing. A legal logbook does not mean 200-and-one-half approaches are wise for you; set reasonable minimums above the legal ones, at least until you have regained full IFR proficiency.

Making yourself legally current can be either time well spent or a mere exercise in order to satisfy the FARs. Because your life and those of others in the IFR system may depend on your proficiency, resolve to gain and maintain all possible skills, not just go through the motions. With time and practice, you can once again be a good instrument pilot.

C H A P T E R
F O R T Y - N I N E

SPLASHING IT ON

Like many aspects of aviation, the field of seaplane flying is deceptively easy to enter. Learning to fly a waterplane isn't difficult; learning to fly one well, however, takes time and dedication. If you're the sort of person who is satisfied with any technique that works well enough to avoid an accident, you probably can pick up a weekend water rating and say "Is that all there is to it?"

If, on the other hand, you like to feel you're the complete master of a machine, dissatisfied with anything less than predictable perfection, you'll still be unhappy with yourself after the quickie water rating is won, and you'll vow to work on the fine touches needed to placate your critical nature. It will pay off in big dividends over the course of a float-flying career, by easing the strain of a seaplane's impact on unyielding water and docks, thereby prolonging the life of floats, attached gear, and airframe.

Seaplane flying is, like soaring or banner towing, pure seat-of-the-pants artistry. It can't be done with instruments or by consulting a page out of a procedures manual. More so than almost any other type of flying, it requires constant use of real-time judgment to gauge the appropriate response. No two days on the lake are ever exactly alike, and thus no single set of actions will work in all cases. You evaluate, you decide, you see what happens, and if your first attempt doesn't work, you throw it away and start over.

SEAPLANE SCHOOL

For the absolute novice, the first decision is where to go to get your single-engine sea rating. There is no written exam to be passed, so you have only to acquire enough skill to satisfy the FAA's practical test standards at the hands of a designated seaplane examiner, usually on staff at the school. Advertisements in the back of the major aviation magazines are a place to start. If at all possible, however, visit the facility in order to look around the place, rather than make a commitment for a $1,000 training course over the telephone. You might find the facility offers down-at-the-heels aircraft, limited water access, and a set of docks better suited for training a Labrador retriever to fetch birds.

Don't expect seaplane bases to look like airports, even in those rare places where both coexist side-by-side. Casual clothing is the uniform of the day. Airplanes used on water tend to accumulate a few dings and scrapes. And the pilot's lounge might be limited to a stump on the shoreline. You'll likely be asked to help with mooring and turning and fueling aircraft, which can be a team effort, particularly when the wind is wrong. In many cases, the seaplane base is part of a marina complex, giving the whole operation a decidedly nautical flavor.

Don't ask to see an instructor's "seaplane CFI" rating unless you want to generate laughter; there is no specific instructor rating for waterplanes. If the instructor is rated for airplanes, he is presumed to be capable of adapting his teaching skills to the waterplanes. So, if his pilot's certificate has a line reading "airplane, single-engine, land and sea" and the CFI ticket says "airplane," the combination is all that's necessary.

DIFFERENCES

With only minor differences, once a seaplane is out of the water, it flies the same as a landplane. The mass and keel effect of the floats hanging under the fuselage give it a slightly stiffer feel, damping control response. Drag, as one might expect, is noticeably increased, slowing the cruise speed. It is especially discernible in the landing approach, where extra power is needed to reduce the sink rate and additional airspeed is carried to prolong the landing flare.

Skimming along on the step just prior to liftoff, this Bellanca Citabria on Edo 2000 floats is in the proper planing attitude to minimize water drag and prevent porpoising.

Tiny tremors run through the airframe as one plows through turbulence, which are tuning-fork vibrations set up by the float rigging. Extra springs may be added to rudder circuits, and the landplane's utility category certification might be demoted to nonaerobatic normal category. Some installations require one or more rear seats to be removed, so pilots won't be tempted to overload an airplane already burdened with 300 or more pounds of extra weight.

Special techniques of takeoff, landing, taxiing, and parking are the essence of seaplane flying, and require 10 hours or more of training for the transitioning landplane pilot. Two factors overlying all seaplane operations are (1) the lack of brakes and (2) the lack of shock absorption in the landing gear. Steering, through the use of water rudders on the back of the floats or fuselage, is easily overpowered by strong winds, requiring knowledge of sailing technique to maneuver in such conditions. Certain things, such as rough water, docking with a following wind, or night landings, are simply avoided.

PREFLIGHT

Preflight is made more complicated by the difficulty of walking around the airplane. If it is moored at dockside, rather than at a buoy, it might be possible to rotate it for inspection, using the ropes usually left dangling from bow cleats or wing tie-downs. Extra hands can save the embarrassment of having to swim frantically for a drifting airplane. The fuel caps are difficult to reach on high-wing airplanes, yet low wings are subject to docking damage.

Make sure the floats are not partially filled with water; pump each compartment through bilge pump openings because water always accumulates in routine operation. In a hull-type amphibian leaving from land, make sure all the drain plugs are in. Check the free and correct routing of the water rudder control and retraction cables and the general condition of the floats and their attaching struts and spreaders. Oil dipsticks might read differently than when the aircraft is on wheels; be sure the engine has the correct one.

TAXI

Taxiing takes place in three basic gaits: (1) Slow-speed, displacement taxiing in a level attitude with the floats creating little or no planning action; (2) the nose-high, plow mode with full throttle and aft elevator, used only to transition to planing speed; (3) a high-speed step taxi at planing speed, when power is reduced to prevent a takeoff but not so far as to allow the floats to fall off the step into displacement mode.

Displacement taxiing is used for close-quarter maneuvering and rough water. Even with the throttle at idle, enough thrust is generated by the slowly turning propeller to move the aircraft forward in calm winds. Seaplane engines typically are tuned to idle as slowly as possible, and quick dependable starting is highly desirable when the airplane is cast off and drifting away from the dock. The water rudders are lowered for displacement taxiing; some small floatplanes need only one water rudder.

To progress to high-speed taxiing on the floats' step, the water rudders are raised, the stick is held full aft, and maximum power is applied, causing the nose to rise and minimizing spray damage to the wildly flailing propeller blades. As speed is gained, the bow wave moves aft along the float. As it does so, back pressure on the stick is gradually released and the airplane soon levels out in a planing attitude, rapidly accelerating into a 40- to 50-mph pace. Power is reduced to an amount that will keep the airplane on the step but not allow liftoff speed to be attained.

At this point, the student begins to learn the attitude control needed to balance an ungainly floatplane on the two patches of water supporting it while it is on the step. Too much back stick causes the heels of the floats to dig in, increasing water drag and requiring more power than necessary. Allowing the nose to come down too far induces a porpoising action, potentially destructive if unchecked, because it usually increases in severity. A slightly higher nose attitude will usually stop the porpoising, although a student occasionally might have to chop the throttle and return to displacement mode. In a short time, one learns where the shoreline should be placed in the windshield for optimum step performance.

Step taxiing increases the pounding of incompressible water on the float hulls that is transferred directly to the fuselage, so its use is limited to relatively smooth water conditions. Step turns require the use of aileron to minimize the tendency of the outside float to bury, which is increased by a turn from downwind to upwind. A constant watch must be maintained for shoals, debris, and the ever-present boat wakes, which might require a turn to cross at a 45° angle to avoid tipping. If in doubt about conditions ahead, come off the step into displacement taxi.

TAKEOFF

Takeoffs are simply a continuation of the step taxi to a slightly higher speed. Gaining this speed is not always easy, however; water drag increases rapidly as the speed goes up, and if the airplane is heavily loaded or the water is smooth (a choppy surface helps air to enter the area under the keels), terminal velocity can be reached short of liftoff speed. Takeoff runs are performed with the maximum-lift flap setting to allow liftoff at as slow an airspeed as possible; raising the wing to a higher angle of attack will only bury the heels of the floats, slowing the airplane down. Rolling in full aileron to lift one float out of the water might cut drag enough to permit acceleration.

Getting a heavy seaplane out of sticky water is a delicate art; takeoffs are sometimes measured in miles.

LANDING

Landing requires planning and preparation; choosing the site involves knowledge of local restrictions (seaplane operations are prohibited on many bodies of water), an assessment of the sea state to determine wind direction and speed, and a check for floating or submerged debris. Old-timers say, "Never land any farther from the shore than you want to swim" just in case you overlook a snag.

A standard rectangular pattern is usually flown to give the pilot a good chance to observe boat traffic and drift. A few knots of extra speed improve forward visibility and increase the time available to adjust touchdown attitude; some power might be carried through the flare for similar purposes. Because the runout distance will be quite short, prolonging the touchdown with extra approach speed is of little consequence.

It is important to touch down in a slightly nose-high attitude, ideally placing the steps and heels of the floats in the water simultaneously. Digging in nose-low induces porpoising, best handled by power application and another, more nose-high, touchdown. Once contact is made, deceleration is

After grounding gently during a nose-in beaching, this Citabria has now been turned around and is secured heels-first on the smooth gravel shore. It will be pushed free for departure, ready for starting as soon as the pilot reaches the cockpit.

rapid and the airplane will quickly fall off the step and revert to an expensive catamaran, if step taxi power is not applied. Additional power and partial flap retraction convert the landing into a splash-and-go.

When the basic skills of takeoff and landing on the water are conquered, the student must learn to handle unusual conditions, like rough water (minimum speed at touchdown, look for sheltered water) or ultrasmooth, "glassy" water that becomes invisible on short final (set up a 100- to 200-fpm sink rate and let the airplane land itself).

DOCKING

Docking and sailing are the marks of a pilot who has made the successful transition into a boater. The airplane wants to be a weathervane, yet the location of the dock inevitably requires some correction to get the closing angle and wind drift mastered. By placing the controls in different directions, opening a cabin door, lowering or raising water rudders, and adding or reducing power, a pilot can vary the airplane's direction of travel even to artfully back into a docking position. Students soon learn why that paddle is carried under the seat; the engine is cut at a safe distance from the dock to prevent propeller strikes and to make a quick exit to attach a mooring line unless handlers are present.

The better schools provide instruction in ramping and beaching the aircraft, approaching and tying to buoys, and shoreline moorings. A good knowledge of nautical rules is important, because the airplane is considered a boat when it is on the water and must adhere to the same rules as any other water-borne vehicle.

Once mastered, seaplane flying is just about the greatest fun it is legally possible to have in or out of an airplane. The indescribable thrill of the spank, spank, spank of floats or a hull touching down on an open lake, the surge of acceleration as the keels break free of the surface on takeoff, the quiet of an isolated gravel beach where a private picnic awaits—these are secrets kept only by the water flier.

SOARING SPIRITS

To truly master the aviation environment, according to glider pilots, you have to experience soaring flight, the birdlike art of gaining altitude by riding upswelling currents of air, instead of pushing a throttle forward. Therefore, if you really want to revert to the basics of heavier-than-air flight, there is no substitute for a glider rating. Gaining the line that reads "glider" on your pilot's certificate is relatively uncomplicated, and yet it is a challenge that will require mental and physical effort.

When adding a glider rating, it is not necessary to take the glider knowledge exam, but it will be important to gain a thorough knowledge of basic aerodynamics, meteorology relating to gliders, and the hazards of ridge, wave, and cumulus soaring, using information found in FAA publications and the Soaring Society of America's *Soaring Flight Manual*. We also recommend Carle Conway's *The Joy of Soaring* as a training text.

Once the written word is in hand, it's time to seriously enter the world of glider flying. Soaring schools, where the training for a glider rating can be obtained, are usually found near the mountain ridges of the Appalachians and Rockies or thermal-based soaring centers in the plains or deserts. Wherever a source of lift can be found, somebody will be flying gliders.

Because of the back-to-basics nature of glider flying, and its limited capability for routine cross-country flight, the age and medical standards are somewhat reduced compared to that established for powered flight. It is possible to solo in a glider at the tender age of 14, although a candidate must be 16 to receive the private pilot's certificate with a glider rating. A medical certificate is not required, although it is not true that you can fly gliders regardless of your physical condition; in lieu of taking a third-class medical exam, you can sign a statement when you apply for the rating declaring that you have no known medical defect that would prevent you from safely operating a glider. Should you then try to fly with your arm in a cast, or whatever, you will still be in violation of the FARs by not being in compliance with your statement.

WHAT ABOUT MOTOR GLIDERS?

Some gliders are equipped with small engines to allow self-launching until reaching an area of lift, where the engine is shut down and, in some cases, stowed to cut drag or the propeller feathered. Although powered aircraft, these motor gliders can be flown by glider pilots who have a logbook endorsement for self-launching method, complimenting the aero-tow launch method for which they were probably originally endorsed. In fact, a glider rating is *required* if the aircraft is certificated under motor glider rules. For the purposes of our discussion, however, we will describe the training as it is normally conducted in nonpowered gliders.

SIMPLY STRONG

The aircraft will seem basic, even crude, when you first encounter it. Two-place training gliders, like the typical Schweizer 2-33, are devoid of upholstery other than in the seats. Floorboards are just that—varnished plywood decks—and the instrument panel has only an airspeed indicator, altimeter, and compass, along with a variometer, which is a very sensitive vertical speed indicator.

To control glidepath accurately and prevent floating during landing, gliders are equipped with spoilers like these, deployed here to reduce lift.

One of the strangest, yet most important, "instruments" will be a tuft of yarn taped to the nose or canopy, to aid the pilot in detecting even small amounts of skid or slip. Gliders must be built as light in weight as possible, of course, to enable them to soar in areas of weak lift. And yet, their structure is deceptively strong; when gross weight is limited to 1,000 pounds and redline airspeed is only 90 mph, there is less need for massive structure.

Preflight inspection of a glider is uncomplicated; no fuel and oil to check, no propeller to feel for nicks, and there's usually only one tire. However, special attention must be paid to the quick-assembly connections, incorporated into the structure of most gliders so they can be partially disassembled for trailer transport. It would be embarrassing to have a wing strut part company with the aircraft shortly after liftoff, so the bolts and retaining pins that hold the structure in place must be carefully checked.

The glider's controls are roughly similar to those of powered aircraft, with the exception of the spoilers or dive brakes that are opened to increase drag and spoil lift. These are needed to overcome the floating inherent in a glider trying to land and to control the angle of descent. Before each flight, the tow hitch's release mechanism is tested, assuring that the glider will disconnect from the towplane when the release is pulled.

TEAM TIME

One of the biggest departures from powered flight is the team nature of glider operation. Unless you are flying a motor glider, or self-launching sailplane, a crew of three or more individuals will be needed to assist in your launch, which automatically obligates you to help them when it's their turn to fly. Gliders on the ground are as ungainly as a fish out of water, so they must be walked to their takeoff position by two persons balancing the craft on its single tire. Once lined up, someone must help test the tow release by pulling on the rope while you work the release knob, and a volunteer must be summoned to hold up a wing while running alongside during the launch until a bit of airspeed is gained. And of course a tow pilot or winch operator must be available.

In most cases, an airplane tow will be used to pull the glider up to a starting altitude for its unpowered flight. Other methods that might be employed are an auto tow, using a long rope behind a vehicle accelerating down a runway, or a winch tow, which launches the glider by winding a cable around a motor-driven drum. Aero tow, however, offers the advantages of an unlimited choice of release altitudes or locations, subject to ability to pay; therefore most U.S. soaring schools launch with aircraft tugs like a Piper Super Cub or a converted agricultural airplane.

FIRST FLIGHT

The first flight in glider instruction (perhaps only 15 to 20 minutes in length, if no lift is found after releasing from the tow rope), will be spent in learning what to expect on takeoff and during the tow. To launch, the slack in the tow rope will be taken up slowly and the towplane pilot will then begin her takeoff at your signal, indicated by fanning the rudder. Aileron control will be available in a hundred feet or so, and the glider will lift off in a few seconds, at around 40 mph. The towplane, however, will still be struggling down the runway on its wheels, and it is imperative that the glider pilot stay close to the runway; if the glider climbs more than a few feet, the towplane might not be able to get its tail down to lift off. Transitioning "power pilots" will often induce porpoising upon liftoff, because of the sensitive elevators of the glider.

On tow, the glider pilot must learn how to keep the tow rope taut, and how to hold position behind the towplane, either in a low slot beneath the wake turbulence spilling off the towplane's wings, or in a "high-tow" location, above the wake. Much will be made of an exercise known as "boxing the wake," or slipping to one side of the towplane, rising or descending to reach the other tow position, then slipping back into center by opposing stick and rudder. Moving back and forth from one tow position to the other by simply passing through the wake is not hazardous, however.

FREE FLIGHT

At a prearranged altitude the release knob is pulled and the glider is nosed up and away to the right, while the tow pilot slides down and to the left. The strong tail-heavy feel encountered during a powered tow dissipates into the neutral trim of gliding flight, as airspeed deteriorates to the minimum-sink value, a speed just above the onset of a stall that expends altitude as slowly as possible, usually 3 feet per second or less.

Control feel will be different than anything you might have experienced in powered flight, because of the slow airspeed and the large ailerons at the ends of long wings. Rudder will be needed to offset the adverse yaw of the big ailerons, roll response will be ponderous, and the elevators will be light and lively.

Best glide speed, which produces the greatest forward distance for the least altitude expended, will be demonstrated. It is somewhat faster than the absolute minimum-sink speed, and is used for straight-line flying toward an objective, rather than loitering. Your instructor will have you open the spoilers or dive brakes to check their effect on the rate of sink, which will double or triple with the boards extended from the wings. Slow flight and stall practice, perhaps with some spin recoveries thrown in, follow the patterns seen in powered flight, except that gravity must be used to regain airspeed, with no help from an engine.

LANDING

All too soon the landing becomes necessary, and the glider must be worked into the traffic pattern. If no lift can be found, it is not unusual to "beat the towplane down!" As the surplus of altitude dwindles, the glider is flown closer to the airport, hopefully maintaining an upwind location to more easily reach the runway. The downwind side of the field at low altitude is definitely not the place to be in a glider. Key altitudes are established for traffic pattern entry, abeam of touchdown zone and turning to base, and the pattern size is adjusted to meet these target altitudes.

Speed will be increased for the approach, to avoid any chance of a stall and to keep some energy in reserve for downdrafts on the approach. Spoilers are also opened partially to steepen the glide angle. If the glider gets too low the pilot can then close the spoilers, similar to a powered pilot adding throttle in order to make the runway.

With practice, a glider pilot can land within inches of a chosen spot and can brake the aircraft to roll right up to a parking spot. New pilots landing long, however, should have to work off their mistakes by helping other pilots drag their glider back from its long landing to a takeoff position.

Your logbook endorsement for solo, aero-tow launch, can be granted after only a few flights, but much practice will be needed before the FIG (flight instructor, glider) will recommend you for the checkride. The regulatory requirements are quite low, with as few as 3 hours of dual and 10 solo flights to qualify for the rating, but most transitioning pilots will need more than the minimum number.

CERTIFICATION

The practical test will cover emergencies, such as a broken tow rope or a towplane power failure, the usual steep turns, stalls, and landings, and changing positions on tow. When the rating is attained, you will still have much to learn, particularly how to locate lift and extend flight time by soaring to higher than tow release altitude. Distances flown, duration of flight, and altitude gained are goals to meet for the coveted SSA badges. The 5-hour flight for the silver duration badge is an accomplishment in flat country, and the distance goals might be a challenge in ridge or wave soaring areas.

As you continue exploring the world of silent flight, you might want to go to soaring contests or learn to fly the towplane. But the glider pilot primarily enjoys the challenge of besting gravity by his or her personal efforts, using skill to find lift and control the flight path. Certificated glider pilots sometimes look upon powered flight with disdain, just as sailboat skippers avoid meeting powerboat drivers. Remember, though, you started your flight behind a powered aircraft.

C H A P T E R
F I F T Y - O N E

UPLIFTING EXPERIENCE

Fixed-wing pilots, like most of the population, are fascinated by VTOL (vertical takeoff and landing) flight. To a pilot trained to avoid flying too slowly at all costs, the ability to stop motionless while still in the air is amazing, as is the capacity to take off and land from any spot large enough to contain the aircraft's landing gear, rather than a half-mile of pavement.

Helicopters are currently the only practical means of performing a job requiring hovering flight; the new "powered lift" category of pilot license awaits the development of personal Harrier jets. Although their utility is considerably hampered by the public's perception of them as noisy and dangerous, effectively preventing them from landing at other than approved sites except in emergencies, more and more helicopters are being pressed into service for door-to-door personal and business travel in areas where landings are approved. However, heliports are easier to establish in rural or industrial areas than in the residential or commercial sections of a city, so the idea of lifting off from a backyard and flying down to the office remains impractical.

For the fixed-wing pilot, the addition of "rotorcraft, helicopter" to a certificate requires considerable retraining to gain the special skills needed. Although some of the academic items you've already learned will be helpful, such as techniques of cross-country flying, working with air traffic control, the basics of meteorology and the Federal Aviation Regulations, many of the concepts will be totally foreign. Part 61.109(c) of the FARs requires only 20 hours of dual instruction and 10 hours of solo time in order to be eligible for the private helicopter license, but most transitioning fixed-wing pilots will need at least 50 hours of total helicopter time before they will be qualified to take the checkride.

The search for a good helicopter flight school can begin with any large aviation training center where a full range of ratings is offered; rotary-wing training is a common adjunct to the fixed-wing program at these schools. If not, however, don't assume that rotary-wing lessons aren't available in your area. There isn't a lot of interaction between the fixed-wing and rotary-wing communities, so you won't necessarily get complete information from an FBO (fixed-base operator) catering strictly to the fixed-wing trade. A check of the yellow pages under "helicopter services" will usually start you in the right direction, and the nearest FAA district office will have a list of schools in its area.

Because of the maintenance-intensive complexity of the aircraft and the high insurance premiums, the rental rates for training-type helicopters will be two or three times that of a fixed-wing aircraft of similar payload. Be prepared to spend considerably more to add the helicopter rating than you originally spent to gain the fixed-wing license, and expect your lack of experience to yield a steep insurance rate after initially receiving the license. The rewards are worth the trouble, however, because helicopter flying is different, and a lot of just plain fun.

What's different about helicopter flying? Helicopters fly low and generally possess wide-open visibility, so the scenery out in front of the bubble is spectacular and full of seldom-seen details. Weather requirements for helicopters are centered less on ceilings and precipitation and more on flight visibility and winds. Helicopters can operate in visibility as low as a half mile with reasonable safety, because of their ability to slow down and avoid obstructions, or even land, if necessary. Wind is a consideration not only for its effect on range, but also for its ability to use a particular landing site if obstructions permit only one approach and departure lane.

During preflight, a fixed-wing aircraft walkaround involves only one rotating part (the engine and its attached propeller), whereas a helicopter's inspection covers many rotating shafts, bearings, drives, and components. Vibration—inherent in a rotary-wing aircraft with its host of moving parts—extracts a certain toll on the structure, which must be built as lightly as possible to allow a reasonable payload to be lifted by the power available.

The normal light helicopter configuration involves a main rotor with two, three, or four blades and a smaller tail, or antitorque, rotor. These rotors will be driven by a reduction drive rather than directly by the engine, in order to achieve a proper rpm range for the rotating airfoil. A clutching arrangement to release the rotors from the engine will be necessary, allowing them to windmill in the event of engine failure. Reduction drives can be gearcases or belt-and-pulley transmissions, or a combination of both.

Control of the helicopter's flight path involves tilting the main rotor disk, so that part of the downward thrust is angled to one side for forward, sideward, or backward flight. To do this, the angle of attack of each main rotor blade is increased as it approaches the desired side of the disk, adding thrust to tilt the plane of rotation. This is termed "cyclic" control, because the blade's angle cycles during its rotation, and is determined by pressure on the control stick between the pilot's knees.

The amount of the lift produced by the main rotor can be increased or decreased by either adding "airspeed" (increasing rpm) or increasing "angle of attack" (pitch angle of all the blades). The latter method, termed the "collective" control, is performed by raising and lowering a lever beside the pilot's left leg, lifting it up to increase blade angle and, therefore, lift, or depressing it to allow the helicopter to descend.

The helicopter's equivalent to an airplane's critical need for airspeed is rotor rpm; if the rotor rpm gets too slow the helicopter will begin to settle uncontrollably and the blades can bend upward from the load. Lowering collective pitch is the only alternative unless more power is available. Just as with a fixed-wing aircraft, power must match the requirements of the desired maneuver; insufficient power will result in a loss of "airspeed" (rotor rpm) as the maneuver is attempted. Throttle control is incorporated into the collective lever, usually with a motorcycle-type rotating grip; some helicopters add power automatically as the collective is raised to ease the pilot's correlation of the two controls.

Nothing is ever as simple as it seems, of course. The Newtonian principles of action and reaction determine that the helicopter's body will tend to be rotated by the engine in a direction opposite to that of the main rotor, and this must be balanced by the sideways thrust of the antitorque rotor at the end of the tailboom. The torque is not constant, however, varying with the power being pulled from the engine, wind direction, and forward speed. Tail rotor pitch is adjusted through the "rudder pedals" ahead of the pilot. Adjusting the outflow of fan-driven air through curved louvers in a hollow tailboom is an alternative method of antitorque control.

As might be expected, coordination of cyclic, collective, throttle, and antitorque controls keeps the pilot's hands and feet fairly busy, so helicopter

After lifting off into a hover, the pilot of this big Sikorsky S-61 is preparing to transition into forward flight to pick up translational lift so he can climb more strongly.

cockpits are simplified as much as possible to allow the pilot to keep his or her hands on the controls. Buttons on the cyclic and collective are often used to activate the microphone, trim the cyclic, turn on lights, even change radio channels.

The preferred hand for activating other switches and controls is the left hand, allowing the right hand to remain on the more-critical cyclic while the collective is held in place by adjustable friction drag. Thus, the pilot-in-command usually sits on the right side of the cockpit, rather than the left, to make it easier to reach the instrument panel and avionics with the left hand.

The transition curriculum will begin with a single control, the cyclic, then the antitorque pedals are added, and finally the collective and throttle will be introduced. These new attitudes and control inputs are usually practiced at a hover, although they can be initially demonstrated in straight-and-level flight, allowing the fixed-wing pilot to feel out the helicopter's control response in the most stable regime of flight. As forward speed is reduced, however, the new pilot's inputs will begin to slip behind the helicopter's movements until he or she learns to anticipate what will be needed.

Hovering practice is frustrating at first; the object is to maintain height at a few feet above the ground while remaining motionless over a spot. Instead, the machine will dart around the spot ceaselessly, until the instructor halts it with seeming ease.

The next task might be a forward flight transition from the hover, flying up and down a runway, followed by climbs and descents and then

The pilot of a side-by-side helicopter occupies the right seat, rather than the left seat typical of a fixed-wing aircraft, in order to have the less-busy left hand available to operate the center console's controls.

approaches to a hover. Pedal turns, precision hover practice, and emergencies must all be mastered before solo.

Once basic helicopter control is mastered, the handling of power or transmission failures is addressed. The helicopter can "glide" with no power, by releasing the engine's drag from the rotors and allowing the relative wind to turn the main rotor. This windmilling needed to maintain critical rotor rpm is called *autorotation,* and is entered by lowering the collective to the bottom of its travel. If the proper airspeed is maintained the rotor rpm will stabilize at a value that assures controlled descending flight, and when the ground is reached a flare is made to reduce forward speed to zero. By doing this, enough energy will be transferred into the rotating mass of the rotor to allow the collective to be pulled upward to arrest the sink rate for a normal touchdown. To the fixed-wing pilot, the angle and rate of descent are a bit unnerving; typical descent rates are three times those of a fixed-wing glide and the glide angle might be only 4:1. It is at this time that the transitioning pilot discovers that the windows down by his feet are for aiming at the touchdown spot during autorotation.

Precision control of the helicopter is practiced by landing and lifting off in confined areas, frequently required when operating near wires and buildings, slope and pinnacle landings, and rapid deceleration from forward flight to a hover. Hazards to be avoided are covered: settling with power during a vertical descent; operating in the "dead man's curve" of the height/velocity envelope, within which a successful autorotation would not be possible; running out of aft cyclic when attempting to come to a downwind hover; rollover after touching down on a sloping surface; and attempting to hover out of ground effect with a high density altitude.

Loading of the helicopter is somewhat more critical than the average airplane, because the CG range is more limited. Gross weight might have to be adjusted for the density altitude conditions, or a running takeoff performed if the ship has wheeled landing gear. The maximum power requirement normally encountered during rotary-wing flight occurs while bringing the helicopter to a hover. Once forward flight is begun and the rotor picks up "translational lift" from a relative wind of 20 knots or more the helicopter will climb with the same power used to hover and can level out to fly at reduced power. For this reason, helicopters seldom lift off or descend vertically; a departure or approach along an obstruction-free lane in forward flight requires much less power from the engine and preserves autorotation capability.

When the checkride is scheduled, the examiner will ask to see most of the maneuvers included in the training course, including autorotations and precision landings, and will judge your ability to be a safe helicopter pilot by the same criteria used in most other checkrides; control of the aircraft must never be seriously in doubt and any situation requiring the examiner to assume command will be disqualifying. If your instructor has covered all the curriculum you'll have no trouble passing the flight test, and then you can proudly wear the added line on your license: "Rotorcraft, Helicopter."

MULTIENGINE RATING

A multiengine rating is one of the most popular additions to a pilot's certificate. After the checkride, new multiengine graduates delight in telling
their envious single-engine buddies how painless it was and what a gas it is
to fly a big airplane with two engines (probably a huge Piper Seminole).

Although it is true that most experienced single-engine pilots can
complete the multiengine transition training with a minimum of fuss, the
matter of competence shouldn't be taken lightly. Because operating costs
of light twins are several times those of a Cessna 152, every effort is
made to turn out a rated pilot in a minimum of air time. After all, just 10
hours of twin-engine dual instruction can cost half the price of a private
ticket. All of which tends to make for legal—but barely competent—new
multiengine pilots. The operator's insurance companies know this, and in
most cases the FBO can't rent multiengine airplanes to his own graduates until they somehow acquire 25 hours or more of multiengine time in
their logbooks.

The shortest road to a safe twin transition starts on the ground; book
learning is essential to safe operation and is more cost-effective than time
spent in the cockpit. Several hours of study are required for each precious
hour in the air. For your pocketbook's and your widow's or widower's sake,
all the facts shouldn't be learned behind the wheel.

ARE TWINS SAFER?

But why all this talk of accidents and careful training? Aren't two engines
supposed to be safer than one? They are, but only in the hands of a properly
trained pilot. Nearly all twin-engine aircraft are complex, heavy machines.
They move fast and require their pilots to think a bit faster than they
move. Like helicopters, twins can safely expand a pilot's field of operation
when operated within specific limitations. Learning how to define these
limitations is one of the main points of multiengine training.

It is best to approach multiengine transition after a background of flying in high-performance airplanes, gaining familiarization with retractable landing gear and controllable propellers, as well as with the slow-speed handling of a heavily laden wing.

Even before setting foot in the twin-engine aircraft, it's ground school time. To understand how to operate with a twin's limitations you must learn why the twin has limitations in the first place. The study must include theory of twin-engine aerodynamics, normally in terms of conventional twins with wing-mounted engines turning clockwise as seen from the rear. Centerline-thrust twins, such as Cessna's Skymaster, are a simpler breed, and twins with the right engine turning counterclockwise, such as Piper's Seneca, are also special cases.

BASIC TWIN THEORY

With both engines running, a twin is just another airplane, albeit a heavy, complex one. When one engine's power is lost and the aircraft is forced to carry on with the remaining thrust—ah, *that's* when things get interesting. In these circumstances it becomes a very poor flying machine indeed, even deadly if not managed properly. If density altitude is too high, the remaining engine's available horsepower may not be sufficient to maintain level flight, particularly with the aircraft in takeoff or landing configuration.

Here's why: Suppose, just to make it simple, that a certain aircraft needs 200 horsepower to maintain level flight in its minimum-drag condition. With two 250-horsepower engines, this means 300 excess horsepower are available to produce a zippy climb rate or fast cruise. Cage one engine, however, and only 50 excess horsepower will be available, and only under the best of conditions will the airplane be capable of climbing. The quoted single-engine service ceiling of the twin assumes a "clean" airplane (gear and flaps up) with the dead engine feathered and standard atmospheric conditions. At less than gross weight it may be bettered a bit, but it won't be approachable with everything hanging down, particularly when going around or when taking off at a high-altitude airport. Under these conditions, when one engine fails, the twin merely becomes a noisy glider.

The windmilling prop produces a tremendous amount of drag when an engine is lost. Healthy amounts of rudder displacement are required to maintain straight flight at slow speeds and, as speed decreases, a point is reached at which a turn into the dead engine will start despite full rudder to the contrary. The speed where this loss of control occurs is the dreaded V_{mc} or minimum control velocity. The V_{mc} is highest at lower-density altitudes and, to make an effective V_{mc} test, many training demonstrations were conducted with insufficient altitude to recover when the airplane became unmanageable, with tragic results. There is no point in continuing V_{mc} demonstrations beyond the moment that a roll or turn begins. Certainly plenty of altitude—3,000 feet or more—should be under the aircraft when

The sight of a feathered propeller means increased workload and quick decisions for the pilot. The multiengine airplane may have enough performance to continue in level flight, or it may have only extended gliding range under some conditions.

investigating V_{mc}. If the demonstration is prolonged the aircraft may stall, which means a spin is probable, and twins aren't even designed to spin on two engines, much less one.

Much discussion centers around the "critical engine." This refers to the engine the pilot would *least* want to lose, because performance is poorer with this engine shut down than with its opposite number. The left engine is normally the critical engine in American-designed airplanes. The reason for the right engine's poorer showing is P-factor, which causes an airplane to yaw left in a climb unless right rudder is applied. At high angles of attack the propeller disk is tilted to the onrushing air. The descending blade, on the right-hand side of the disk as viewed from behind, is operating at a higher angle of attack as it turns than the ascending blade on the left. Thus the right-hand blade produces greater thrust. The effective center of the prop's thrust is no longer at the crankshaft, but moves to the right side of the propeller disk. This produces the yaw moment called P-factor, particularly significant in twins.

THE TWIN'S *V* SPEEDS

On a twin, the lowest V_{mc} speed is limited by the rudder control available to combat the yaw of single-engine flight. Because the center of thrust from a propeller operating at low airspeeds is displaced to the right, it follows that yaw from the right engine's asymmetrical thrust is made even worse by the P-factor. The left engine's yaw, on the other hand, is lessened by the P-factor shifting thrust to the inboard, rather than the outboard, side of the propeller disk. Thus all other things being equal, the airplane will fly with less drag and down to a lower V_{mc} on the left engine, making it the critical one to lose. Piper's Seneca and Cessna's Crusader, for example, are fitted with counter-rotating engines in order to place offcenter thrust from P-factor inboard regardless of which engine loses power. This allows a lower certificated V_{mc} than would otherwise be possible.

Other speeds must be added to the stall and climb speeds a student normally learns in single-engine training. V_{mc} is one, of course, but that speed guarantees only *control*, not continued level flight or climb or even the ability to accelerate. Going on up the scale, one finds the best single-engine angle-of-climb speed, or V_{xse} for short. Holding this speed will bring the steepest possible climb, assuming the airplane has been cleaned up.

Best rate of climb on one engine, or V_{yse}, is a bit faster and offers a flatter climb but a greater rate of ascent. So to achieve true twin-engine safety you must not let the airplane leave the ground before V_{mc} is reached, hopefully quite a bit more; ideally the twin should have a long-enough runway to accelerate almost up to V_{xse}, lose an engine, and still stop on pavement.

This is the reason twin-engine pilots prefer long runways—not because the airplanes require extra length for the takeoff run, but because it is safer in case engine failure on takeoff makes a stop necessary. On short runways, a twin is frequently seen to consume almost the entire paved length before liftoff. "Boy, he barely made it!" exclaims the shocked onlooker. This isn't the case at all. The airplane was capable of flying long before the liftoff point, but the pilot wished to accelerate until V_{mc} or V_{xse} was achieved.

FINALLY, WE FLY

After doing enough ground study to make expensive airborne training useful, it's time to go flying. Preflighting a twin offers no surprises beyond the obvious dual power plants and, usually, a more complex fuel system with more sumps to drain. On board the aircraft, a few minutes should be spent in reviewing the cockpit layout before starting the engines. A multiengine cockpit is "busier" than that of a lighter single-engine airplane, so it's wise to climb in and just sit for a while. Take the operator's manual home and sleep with it. It will probably be thicker and tougher than any you have digested so far in your career, and there is no substitute for knowing the aircraft—especially a new, challenging twin.

Expect to use a checklist in every twin to avoid overlooking a vital assessment or manipulation in proper sequence. Clear the area around the propellers carefully—especially the right prop, which may be difficult to see—before starting the engines. With both engines running and all systems "Go," the aircraft is ready to taxi. The amount of power needed to initially move the heavier airplane may be surprising. New twin pilots like to play inboard and outboard throttles to aid in making taxi turns, but it isn't necessary to carry this to extremes. With nosewheel steering, differential power isn't really needed; generally, retarding throttle on the inside of the turn is the best procedure, rather than "blasting" outside throttle.

After the runup checklist is completed, the takeoff must be handled with a bit more planning than in single-engine airplanes. Remember, your concern is not so much for gaining the speed necessary to get the airplane off the ground as it is for attaining the capability to continue flight if an engine fails. V_{mc} is marked by a red line on the airspeed indicator in newer twins, and V_{yse} is shown by a blue line. Ideally, you should leave the nosewheel on the runway until passing V_{mc} and plan to lift off around V_{xse}, which is just below V_{yse}. If an engine fails before the aircraft reaches V_{mc}, chop the throttles and stop. If an engine is lost after passing through V_{mc}, stop if at all possible, but you may opt to hold V_{xse} carefully and clean up the airplane in an effort to make it fly. Speed in raising the gear and flaps and feathering the propeller on the dead engine is most important because the airplane will likely lose altitude at V_{xse} until the drag is lessened. In any event, do not permit airspeed to drift below V_{mc} during this phase, because full control would no longer be available.

So, the twin is lined up for takeoff and the throttles are eased in smoothly. This time everything keeps running and the aircraft proceeds skyward for a little airwork. The first hour or so of dual will be spent primarily in learning to fly the airplane with both engines operating, through slow-flight, stalls, and other maneuvers. In later flights, single-engine drill will be learned until procedures become second nature.

WHEN TWO BECOME ONE

When one engine fails, the first job is to control the sudden yaw of the airplane into the dead engine. Normal use of rudder and aileron will hold the bird straight, but the pressure may be heavy—the student multiengine pilot can expect a little soreness in her leg muscles after a good single-engine workout. While holding the airplane straight, maximum power is applied and steps are taken to clean up the airplane. This means all power controls are usually shoved full forward, without regard to engine gauges for the moment, and gear and flap controls are placed in the "up" position. This procedure helps the airplane hold its altitude for a few seconds until you can feather the windmilling prop. A glance at the airspeed shows we're still above V_{yse}.

Identification of the dead engine comes next—and it's vital to be absolutely correct. First, which foot is not holding rudder pressure? The dead foot is the dead engine, according to the time-honored cliche of twin training, and it works. The pilot who's holding hard right rudder knows her airplane's left engine is out, but just to be sure, she'll ease back the left throttle and see if engine noise and yaw remain unchanged. They do, so the dead engine has been identified.

Now the prop control is moved back into "feather" position and the dead engine's mixture control goes rearward to "cutoff." At this point, the busy pilot is still hanging onto V_{yse}. Using a 5° bank into the good engine will ease the rudder pressure required for straight flight and reduce drag to assist climb somewhat. Mags, boost pump, alternator, and fuel valve related to the dead engine are carefully turned off at this point.

Having learned to control the twin after losing an engine, and how to make a prompt—but not mechanical—shutdown of the windmilling engine, the rest of the training will seek to polish those skills in simulated emergency situations, such as losing an engine just after going around from a balked landing. At cruise speed an engine failure is no urgent catastrophe, but in full-dirty go-around configuration, survival depends on fast and correct handling of the situation.

Initial practice of emergency procedures is carried out at altitude, where full shutdowns to feather can be safely attempted if desired. Then the training moves down into the traffic pattern for the real thing. There is no substitute for the feel of terra firma rushing by right under your hip pockets. Often a student who is cool and collected at altitude forgets everything she has learned when she sees the asphalt coming up at her at a great rate of speed. Naturally, low-altitude engine failures are only simulated. Instead of allowing the engine to be feathered, the instructor sets up 10 inches or so of manifold pressure—for zero thrust—so power can be instantly restored should the need arise.

Ten hours of dual instruction is an average twin-engine transition time for a very competent pilot, although some have earned the rating in less. No knowledge test is required in advance of the checkride, but the oral exam is thorough, as the examiner makes sure the candidate has done her homework. Do exactly as you have been taught during the in-flight portion of the checkride and, presto, you will be a many-motor driver. If you are instrument rated, you will have to prove your ability to handle an engine failure and make an approach with the hood on in order to gain multi-engine instrument privileges.

As stated earlier, your chances of going out and renting a twin right away are practically nil, so seek every chance to fly copilot or resign yourself to additional dual instruction. If you are fortunate enough to get to fly a twin regularly, make sure you ride at least annually with a competent multiengine instructor to keep sharp on procedures rarely practiced after training. You'll enjoy every minute behind those twin throttles!

ATP CERTIFICATE

A popular joke sheet on flight school bulletin boards states that airline transport pilots are able to leap tall buildings at a single bound, walk on water, and generally excel at all sorts of challenging endeavors. This is not exactly untrue; ATPs can leap tall buildings with the aid of a Boeing 737 and have been known to walk on the water of a frozen hockey rink. In reality, ATPs are just competent professional pilots who have undergone a disciplined course of training to achieve the highest pilot rating bestowed by the FAA.

ATP DEFINED

The airline transport pilot is perhaps the least-understood pilot certificate. It carries only a few extra benefits beyond those of a commercial pilot's certificate with an instrument rating, the primary one being the dubious privilege of giving "instruction in air commerce." Because the ATP written and flight tests demand a high degree of knowledge and skill, the rating also has become a hiring requirement for the most serious flying jobs, which is the reason it is sought after by eager young pilots.

If two equally experienced pilots were to apply for a job, the ATP would probably get the nod over a mere commercial/instrument pilot. Commuter airline captains are now required to be ATP rated, as is the pilot-in-command of an air taxi powered by jet engines or having 11 or more passenger seats. Although there is a single-engine ATP rating, as well as a helicopter ATP certificate, the multiengine ATP is the usual working ticket sought for employment purposes.

The ATP originally was promulgated as an upgrade awarded to airline copilots moving over to the captain's seat. The applicant would be back-grounded in en route and approach procedures, be a current instrument pilot, and be thoroughly familiarized with the aircraft's systems by a transition school. In air carrier training programs, much of the checkride can be performed in a sophisticated simulator.

ON YOUR OWN

Individual pursuit of the ATP, a popular pastime with aspiring airline hopefuls, can be undertaken in as structured a manner as desired. Just about any twin-engine airplane up through bizjets can be used for the checkride, and no actual dual need be logged in preparation. However, it would be foolish to attempt the flight test without getting a lot of practice with an ATP in the right seat, preferably one who is also a practicing multiengine CFI. A formal flight school is always best, especially because the FAA, in addition to FAA inspectors, allows certain designated examiners working with these schools to give the ATP checkride. A typical 30-flight-hour/40-ground-school-hour ATP training course costs nearly half the price of a good automobile, however, and the rating can also be won by simply taking the flight test guide seriously and practicing in a familiar cockpit.

Basic requirements for the ATP include 1,500 hours of pilot time, of which 500 hours must be cross-country flying, 100 hours of night time, and at least 75 hours of instrument flying. The first-class medical exam is not required unless one wishes to exercise the privileges of the rating. No recommendation is necessary to take the knowledge or flight tests. Evidently the pilot is supposed to know when he or she is ready, perhaps seeking the counsel of an ATP before making application.

The knowledge exam is quite exhaustive, requiring a thorough knowledge of Part 121 or Part 135 of the FARs, covering day-to-day commercial operations. Achieving a passing score on such a test might not be of much practical value in general aviation flying, but it does prove that the individual can absorb an amazing amount of trivia if required. The exam consists of 80 multiple-choice questions in the style of all FAA written exams, with a 3-hour time limit.

The oral portion of the practical test is designed to reveal the applicant's knowledge of his or her airplane's systems and limitations. The pilot's operating handbook must be memorized, for all practical purposes. Page-by-page explanation may be required, with "what-if" questions posed at various points. Performance, loading, flight planning—all must be discussed in detail to the inspector's satisfaction before entering the airplane. A thorough preflight inspection is made, during which the applicant is expected to know the function and terminology for each item checked.

FLIGHT TEST

The ATP flight test, by comparison, is nothing more than a highly critical instrument proficiency check taken in a high-performance airplane. The ATP practical test guide lays it all out—each required maneuver, its desired objectives, and acceptable performance. The criterion is generally a

The sight of this regal Boeing 737 brings out the pilot fever in everyone. The occupant of the left seat will be an ATP, the highest rating awarded by the FAA.

commercial pilot level of competence delivered despite such diversions as power plant failures or equipment malfunctions. Altitude limits are generally plus or minus 100 feet, heading errors must be limited to 10°, and airspeed parameters are normally 10 mph. These are not overly generous margins in a high-performance airplane, and the task is often complicated by the need to control speed, altitude, and heading simultaneously while managing an engine shutdown.

The acceptable performance window narrows somewhat during critical moments, as during an instrument approach. MDA tolerance is plus 50 feet, minus zero, for example—not an easy target in low-level turbulence while running out the estimated time to the airport. The circle-to-land maneuver must be completed within the appropriate minimum-visibility radius without exceeding a 30° bank angle, and altitude tolerances are 100 feet above MDA, zero feet below. The manual ILS is to be flown with precision befitting a coupled autopilot, not more than a one-dot deviation right down to the decision height (DH) while maintaining airspeed within 10 knots. Two types of nonprecision approaches must be demonstrated in addition to the ILS.

Engine-out emergencies are almost routine during an ATP checkride. The applicant is expected to maintain control and continue with the appropriate maneuver—be it a takeoff, ILS approach, or missed-approach procedure—while securing a failed engine. Tolerances for error are unchanged with an engine out; the same one-dot deviation on the ILS and 10-knot airspeed control must be maintained. A thorough knowledge of cockpit controls is vital at this time.

The aspiring ATP should expect a heavy workload during the flight check; instruction and practice will emphasize this. One ATP checkride of my acquaintance required a manual landing gear extension during single-engine holding at an NDB, as an example. Flying the aircraft with precision must be almost second nature, allowing part of one's attention to be directed elsewhere—beating out flames or whatever. Quite simply, the ATP is expected to be able to cope with any conceivable emergency at any moment, ensuring the survival of his passengers under the most adverse conditions. It is a serious rating—attainable by anyone willing to undergo stringent preparation but not easily obtained with a week's practice.

If you feel an ATP rating would improve your self-esteem, total up your hours and get a knowledge exam study guide or enroll in a ground school course. The exam results are good for 2 years, giving ample time to practice for the flight test. If you're really serious about precision flying, you can join the ranks of airline transport pilots and learn to walk on water yourself.

INDEX

accidents, 213–216
  fatalities, 213
  flight review vs., 219–223
  hours of flying time vs., rates, 4
  insurance and, 215
  midair collisions, 43–46
  multiengine aircraft and,
    281–282
  notification, reporting,
    preservation, 214
  poor pilot judgment, 155
  serious injury, reporting, 213
  substantial damage, reporting,
    213
  weather vs., rates, 6
aerial photography, 10
aerobatics, 5, 225–228
  aircraft designed for, 227–228
  G loading and stresses, 225
  illegal, 173
  instruction for, 226–228
Aeronautical Information Manual
  (AIM), 48, 153
afterburner, 166
air filters, 57
Air Force Rescue Coordination
  Center, 34
air sickness, 21
air stagnation, 130
air traffic control, 33
  Class B airspace, 69–75
  Class C airspace, 65–68
  grass-strip airports, 53–57
  lack of, at uncontrolled airports,
    47–51
  midair collisions, 43–46
  patterns, 49–51

air traffic control (*Cont.*):
  proper procedures with, 169
  right-of-way, 48
  tower-controlled airports, 59–64
  traffic advisory frequency,
    common, 51
Aircraft Owners and Pilots
  Association (AOPA), 8
airframe:
  icing conditions, 143–144
  postflight inspection, 210–211
airline transport pilots, 287–290
  flight test for, 288–290
  instruction for, 288
Airport/Facility Directory, 7, 61
airports, 7–8
  Class B airspace, 63, 69–75
  Class C airspace, 61, 63, 65–68
  grass-strip, 53–57
  tower-controlled, 59–64
  uncontrolled, 47–51
airworthiness certificate, 189
airworthiness directive, 191
alcohol use, 174
alternate airports, 7, 9
altimeters, 70–71
  encoding, 71
altitude:
  engine performance vs., 125
  fuel consumption vs., 13–14
  limits, FARs for, 23
  low-level flight, 23–27
  minimums, 174
amortization, flying clubs and,
  185
approach procedures, 65–70
  chart showing, 65

approach procedures (*Cont.*):
   clearance, 71
   departures, 67
   landings, 62
   takeoffs, 64
approaches, 46
   Class B airspace, 73–74
   commercial pilot requirements,
      246
   high-performance aircraft, 241
   maximum performance, 246
   midair collision avoidance and,
      44–46
   missed, 262
   NDB, 261
   proper procedure for, 169
   tower-controlled airports, 62
   uncontrolled airports, 49–50
   VOR, 261–262
area navigation (*see* RNAV)
atlas, 7
automatic direction finder (ADF),
      25, 248
automatic terminal information
      service (ATIS), 60, 63–70, 169
autopilots, 262
autorotation, 279

banking, 4
batteries, 138, 211
bill of sale, 189
buying airplanes, 5, 189–192
   airworthiness certificate, 189
   airworthiness directives, 191
   bill of sale, 189
   checkout procedures, 191, 239
   insurance, 190–191
   liens against, 189–190
   local taxes and registration, 190
   logbooks, 189
   operating limits, 189
   radio license, 190
   registration, 189–190
   test flight, 203–205
   weight and balance data, 189
buzzing, 173–174

call signs, 61–62
canyons, 27
carburetor, 57
   cold starts and, 137
   fires in, 138
   icing, 141–143
ceilings, 34, 83, 88
   marginal VFR, 94–95
certified flight instructor (CFI), 5
Cessna, 229
Cessna 185, 234
chandelles, 245–246
charts, 10
   IFR tests, 247–249
   low-altitude, 9
   sectional, 7–9, 25
checklists, 193–199
   in-flight, 198–199
   preflight, 196–197
Civil Air Patrol, 27, 34
Class B airspace, 63, 69–75
Class C airspace, 61, 63, 65, 68
clearances, 62–64, 67
climbs, 4, 10
   constant-speed propellers, 16
   fixed-pitch propellers, 16
   fuel consumption economy
      during, 16
   multiengine aircraft, 284
clouds, 87–88
   flying through, VFR, 100
   flying under/over, 97–98
   fog, 131–134
   haze and, 127
   icing conditions in, 142
   marginal VFR, 94–95
   thunderstorm formations and,
      112–113
cold fronts, 86–87, 89
commercial pilots, 243–246
   flight maneuvers required by,
      245–246
   IFR and, 247
constant-speed propellers, climb
      speeds, 16
control towers, 36, 59–64, 169

controlled airports, 56–64
controls and control surfaces,
    checklist for, 194
costs (see economy)
course deviation indicator (CDI),
    17
cross-country, 5
    IFR flight planning, 252–253
    VFR flight planning, 33
cruise, fuel consumption and, 16
currency, 177–181
    IFR, regaining, 259–262
    preflight preparation, 178–179
    taxi and takeoff, 180

damages, 213
dead reckoning, 25–26
departures, 19–20
    Class B airspace, 74–75
    Class C airspace, 67–68
    noise abatement procedures,
        166–167
    proper procedures for, 165–168
    tower-controlled airports, 63–64
descents, 4, 10, 17
    dump areas, 45
    fuel consumption, 16–17
    midair collision avoidance and,
        44–45
    tower-controlled airports, 61
    VFR in IFR conditions, 99–100
destination airport, 5, 7, 17
directional gyros, 99
docking seaplanes, 268
downdrafts, 113
drift angles, 86
dump areas, 45

economy, 13
    amortization of loans, flying clubs
        and, 185
    flying clubs, 185–188
    IFR rating, 249–250
    rental airplanes, 186
    renting vs. buying airplanes, 5
electric lines, 27

embedded thunderstorms, 130
emergencies, 157–160
    emergency locator transmitter
        (ELT), 33–34, 209
    engine failure, multiengine
        aircraft, 285–286
    equipment for, 21
    forced landings, 34
    transponder code 7700, 101
    VFR in IFR conditions, 100
emergency locator transmitter
    (ELT), 33–34, 209
encoding altimeters, 71
engines:
    altitude vs. performance, 125
    carburetor fires, 138
    carburetor icing, 141–143
    care and maintenance, 138–139
    checklist for, 194
    cold starts, 135–139
    cooling fins and baffles, cleaning,
        124
    flooding, 125, 137–138
    hand-priming, 137
    heat stress and, 123–124
    hot starting, 125
    jump starts, 138
    multiengine aircraft, 281
        failure of critical engine,
            282–286
    new, test flight for, 202–205
    oil leaks, 210
    oil levels, 124
    performance deterioration,
        125–126
    postflight inspection for, 209
    preheating, 135–137
    rebuilt, test flight for, 202–205
    return-to-service check/test
        flight, 201–203
    shutdown, inspection after,
        210–211
    unfamiliar aircraft procedures,
        180–181
estimated time of arrival (ETA),
    10

exhaust gas temperature gauge (EGT), 15

fatalities, 213
Federal Aviation Regulations (FARs):
  altitude limits, 23–24
  commercial pilot requirements, 243–244
  high-performance aircraft, 237
  illegal and unsafe acts, 171–172
  weather briefings and, 85
fires, carburetor, 138
fixed-pitch propellers, climb speed, 16
flight-following service, 33
Flight Guide, 8
flight logs, 10
flight planning, 7–13
  closing, 36
  emergencies vs., 33–34
  filing, 34–35, 257
  IFR, 251–254
  IFR vs. VFR decisions, 255–258
  planning charts, 9
  VFR, 33–36
flight reviews, 5, 219–223
flight service station (FSS), 33–35
  preflight weather briefings, 79–83
  VFR in IFR conditions, help, 100
flying clubs, 185–188
fog, 26, 131–134
  formation of, 131–134
  IFR decisions, 132–133
  VFR flight, 133–134
forecasts, 82
frontal thunderstorms, 113–114
frontal zones, 10, 86, 89, 114
fuel consumption, 7–10
  altitude vs., 13–14
  climbs, 16
  cruise, 16–17
  descents, 17
  leaning mixture and, 14–15

fuel consumption (Cont.):
  logging, 18
  power settings vs., 13
  route planning vs., 15–16

G loading, 225
gliders, 269–274
  certification for, 273
  controls and control surfaces, 271
  free flight, 272–273
  instruction for, 272
  landing, 273
  launching procedures, 271
  powered aircraft vs., 271
  team cooperation for, 271
  towing procedures, 272
  wake turbulence, 272
global positioning system (GPS), 25, 37–39
  approaches, 37, 261
go-arounds, 50, 62
go/no-go judgment, 153–156
grass strips, 53–57
gross weight, 175, 279
groundloop, 231
groundspeed, 10, 25
GUMP check, 241

hand-priming, 137
haste, 157–160
haze, 86–87, 127–130
  embedded thunderstorms, 9, 130
  navigation in, 129
  VFR in IFR conditions, 129–130
headwinds, 10
heat stress, 123–126
helicopters, 275–279
  autorotation, 279
  controls and control surfaces, 276–277
  flight maneuvers for, 278–279
  instruction for, 275
  licensing and ratings, 275
  loading, weight and balance, 279

helicopters (*Cont.*):
  preflight, 276
  transitioning practice, 278–279
  translational lift, 277, 279
high-density traffic, 5
high-performance aircraft, 237–241
  approach, 241
  avionics for, 239
  checkout for, 239–240
  controls for, 240
  FARs concerning, 237–238
  flight maneuvers for, 240
  grandfather rights, 238
  landing, 241
  licenses and ratings, 237
  slow-flight, 240
  stalls, 240
  takeoff, 240–241
  transitioning to, 240
high-pressure systems, 86
high terrain, 9, 27
hot starts, 125

icing, 9, 16, 131, 141–144, 250
illegal acts, 171–175
in-flight checklist, 195–196
instrument flight rules (IFR), 9–10,
    16, 43, 45, 85, 247–250
  autopilots and, 262
  cockpit duties delegation for, 261
  commercial pilots, 243, 247
  costs of, 249–250
  cross-country flight, 250
  currency in, 259–262
  flight planning, 251–254
  flight simulator for, 259
  fog, 133
  haze, 128
  icing conditions and, 143–144
  no-radar procedures, 261
  rating requirements for, 247–248
  thunderstorms and, 113
  VFR vs., 255–258
  weather conditions for, 97–101
  weather vs., 249, 253–254

instrument landing system (ILS),
    248, 261
instruments, checklist for, 194
insurance, 190–191, 215, 221

Jeppesen charts, 9, 31, 249
Jeppesen's JeppGuide, 8
jump starts, 138
jumper cables, 21

landing gear, postflight inspection,
    211
landings:
  Class B airspace, 73
  Class C airspace, 67
  commercial pilot requirements,
    245
  downhill-sloped runway, 57
  emergency, 33
  gliders, 273
  grass strips, 53–57
  high-performance aircraft,
    240–241
  maximum performance, 246
  off-airport, 172–173
  postflight inspection after, 208
  proper procedures for, 169
  seaplanes, 267–268
  snowy runways, 149
  tailwheel aircraft, 233–235
  tower-controlled airports, 60–62
  uncontrolled airports, 50–51
  unfamiliar aircraft procedures,
    181
  uphill-sloped runway, 56
  winds and, 100, 109
launching gliders, 271
lazy eights, 245–246
leaning, fuel consumption vs.,
    14–15
licenses:
  airline transport pilot (ATP),
    287–290
  commercial pilot, 243–246
  currency and, 177

licenses (*Cont.*):
  flight reviews and, 220
  gliders, 273
  helicopters, 275
  high-performance aircraft, 237
  IFR rating, 247–250
  seaplane ratings, 263–264
liens, 190
light system, 45, 211
logbooks, 189
low-altitude en route charts, 31
low-level flight, 23–27, 98
  limitations to, 23–24
  marginal VFR and, 94–95
  navigation, 24–25
  obstructions and hazards, 25–27
  weather conditions, 23
low-pressure systems, 79, 86

marginal conditions, 83, 91–95,
    255
  ceilings, 93, 95
  clouds, 94–95
  visibility, 91, 94
  VOR airways to follow, 93
maximum performance
    approach/landing, 246
midair collisions, 43–46
mileages, 10
military operations areas (MOAs), 9
multiengine aircraft, 281–286
  accident statistics and, 281–282
  failure of critical engine,
      282–283
  P-factor (yaw), 283
  preflight, 284
  takeoff, 285
  V speeds for, 283–284

National Transportation Safety
    Board (NTSB), 213–215, 219
National Weather Service, 80, 83
navigation, 7, 24–25
noise abatement procedures,
    166–168

nondirectional beacon (NDB),
    261–262
notices to airmen, 82, 153

obstructions and hazards:
  grass-strip runways, 55
  low-level flight, 25–27
  towers, 29–32
off-airport landings, 172–173
oil leaks, 210
oil levels, 124
orographic thunderstorms,
    113–114
overloading, 175
oxygen, supplemental, 9, 14

P-factor, 283
parking:
  proper procedures for, 169
  uncontrolled airports, 51
passengers, 4
patterns, traffic (*see* air traffic
    control)
performance, 10, 13–14, 16
Piper, 229–230
postflight inspection, 207–212
  squawk sheets, 208–210
power settings, fuel consumption
    vs., 13–14
precipitation, 82–86
preflight inspection, 4
  customized checklist for, 193–199
  gliders, 271
  helicopters, 276
  multiengine aircraft, 284
  seaplanes, 265
  stress relief during, 162–164
  weather briefing, 79–83
preheating devices, engine, 136–137
pressure-pattern flying, 10
propellers, hand-priming, 137

radar, 33, 69, 75
  no radar, IFR procedures,
      261–262

radio, 33, 61–64, 70
  Class B and Class C airspace,
    65–75
  license for, 190
  postflight inspection for, 209
  proper procedures for, 166
  uncontrolled airports, 51
ratings (*see* licenses)
rebuilt airplanes, test flight,
  203–205
refresher training, 5
registration, 189–190
remain overnight (RON) kit, 11, 20
renting airplanes, 56, 186
repairs:
  postflight inspection for,
    207–212
  squawk sheets, 208–209
restricted areas, 9
return-to-service check, 201–203
right-of-way, 49
RNAV (area navigation), 9, 15,
  248, 261
road atlas, 7
routes, 8–9, 15, 20
runway manners, 50
runways:
  downhill slope, 57
  frozen surfaces, 55
  obstructions and hazards, 55–56
  snow-covered, 147–149
  soft surfaces, 55
  uphill slope, 56

seaplanes, 263–268
  docking and sailing, 268
  landings, 267–268
  landplanes vs., 264–265
  preflight inspection, 265
  takeoff, 266–267
  taxiing, 265–266
search and rescue, 27, 34
sectional chart, 7–8, 25
  Class C airspace indicated on,
    65–66

sectional chart (*Cont.*):
  obstructions listed on, 25–26,
    30–31
separation, 43, 62, 73
shock struts, 57
slope, runway, 56–57
snow, 145–149
soaring (*see* gliders)
Soaring Flight Manual, 269
Soaring Society of America, 269
spins, 171–172
spirals, 245–246
squall lines, 117–121
squawk ident, 71
squawk sheets, 208–210
stagnation, air, 130
stalls, 204, 240
stress management, 161–164
  in-flight relief techniques, 164
  preflight relief, 162–164
survival kits, 21

tailwheel aircraft, 229–235
  center of gravity in, 230–231
  ground handling, 231–232
  groundloop, 231
  landing, 233–235
  takeoff, 233
  taxiing procedures, 232–233
  transitioning to, 230–231
tailwinds, 10
takeoffs:
  checklist for, 194–195
  Class B airspace, 74–75
  Class C airspace, 67–68
  commercial pilot requirements,
    246
  grass-strip runways, 55
  high-performance aircraft, 240
  multiengine aircraft, 285
  seaplanes, 266–267
  snowy runways, 147–148
  tailwheel aircraft, 233
  tower-controlled airports,
    63–64

takeoffs (*Cont.*):
  unfamiliar aircraft procedures,
    180
  winds and, 107–108
taxiing:
  proper procedures for, 165
  seaplanes, 265–266
  snowy runways, 147–148
  tailwheel aircraft, 232–233
  unfamiliar aircraft procedures,
    203
temperature lapse rate, 10
temperatures, 86, 89
  high temperatures/heat stress,
    123–126
tents, 21
terminal areas (Class B), 65, 69–75
  approaches, 70–73
  communicating with tower,
    73–74
  departures, 74–75
  landings, 75
  radar vector assigned, 72
  radio frequencies, 70–71
  separation, 69
  takeoff, 74–75
  touchdown, 73
  traffic advisories, 72–73
terminal radar service area (TRSA),
  69, 75
test flights, 201–205
  new airplanes, 203–205
  rebuilt airplanes, 203–205
  return-to-service check, 201–203
thermals, 89
thunderstorms, 9, 86, 111–115,
  131, 249
  avoidance of, 113–115
  cloud formations for, 111–113
  embedded, haze, 130
  frontal, 111–112
  orographic, 112
  squall lines, 117–121
  turbulence in, 118
  up- and downdrafts, 113

tiedowns, inspection of, 112, 212
toilet facilities, 21
tower-controlled airports, 59–64
  arrival procedures, 60–62
  clearances, 62, 64
  departures, 63–64
  descent, 62
  go-arounds, 62
  landings, 62
  on-ground procedures, 62–63
  takeoffs, 64
towers, 25–27, 29–32
traffic advisory calls, 49
traffic advisory frequency, 51
translational lift, helicopters, 277,
  279
transponders, 64–65, 71
  emergency code 7700, 101
trim, 99
true airspeed (TAS), 10
turbochargers, 14
turbulence, 31, 88–89, 98, 108,
  168, 249
  thunderstorm/squall line, 118
turns, VFR in IFR conditions, 99

uncontrolled airports, 47–51
unfamiliar aircraft, procedures for,
  177–181
Unicom, 48–49, 51, 169
unsafe acts, 172–175
updrafts, 113

vacation planning, 19–22
visibility, 29, 83, 88, 91, 98
  fog, 131–134
  haze and, 127–130
  marginal VFR, 94–95
  midair collision avoidance and,
    44–45
visual flight rules (VFR), 9–10, 16,
  44, 83, 88
  clouds, flying through, 99–100
  clouds, flying under/over, 97–101
  commercial pilots, 243–244

visual flight rules (VFR) (*Cont.*):
  emergency frequency for, 100
  flight plan, 33–36
  fog, 133
  haze, 128
  icing conditions, 141–144
  IFR conditions during, 97–101
  marginal weather, 91–95, 257
VOR, 9, 17, 33, 35, 39, 93–94, 98,
  248, 261
VOR/DME, 261
VTOL aircraft (*see* helicopters)

wake turbulence, 73–74, 272
warm fronts, 86
water, 21, 263
weather, 6, 9–10, 16, 20, 22–23,
  34, 44
  air stagnation, 130
  ceilings, 88
  clouds, 86–87
  FARs concerning, 85
  fog, 131–134
  forecasts, 82, 86
  frontal zones, 26, 29, 114
  haze, 86–87, 127–130
  heat-stress, 123–126
  high-pressure systems, 86
  icing, 141–144
  IFR vs., 250, 254–256
  IFR vs. VFR decisions, 255–258
  low-pressure systems, 86
  marginal conditions, 83, 91–95

weather (*Cont.*):
  minimums and limitations,
    154–155
  national sources for, 80–81
  orographic thunderstorms,
    114
  personal forecasting, 85–89
  pilot's reports on, 168
  precipitation, 86
  preflight briefing, 79–83
  regulations concerning, 154
  snow, 145–149
  squall lines, 117–121
  temperatures, 86, 89
  thunderstorms, 87, 111–115,
    130
  turbulence, 88–89
  visibility, 86–89, 91
  winds, 86–88, 105–109
Weather Channel, 80–81
weight and balance, 19, 189
  helicopters, 279
  overloading, 175
winds, 21, 88, 105–109
  landing, 108–109
  speed/direction indicators,
    105–106
  squall lines, 117
  takeoff, 107–108
  taxiing, 107
  wind speed vs. stalling speed,
    105–106
winds aloft, 23

ABOUT THE AUTHOR

LeRoy Cook calls himself a lifelong aviation student. A columnist and editor for *Private Pilot* magazine for more than 30 years, and a flight instructor for nearly 40 years, he holds FAA Gold Seal certification for single-engine and multiengine airplanes, gliders, and instrument instruction.